YOUR BRAIN IS PLAYING TRICKS ON YOU

YOUR BRAIN IS PLAYING TRICKS ON YOU

How the Brain Shapes Opinions and Perceptions

Albert Moukheiber

Translated by
Anne-Sophie Marie

HERO, AN IMPRINT OF LEGEND TIMES GROUP LTD
51 Gower Street
London WC1E 6HJ
United Kingdom
www.hero-press.com

First published in French as *Votre cerveau vous joue des tours* by
Allary Éditions in 2019
This translation first published by Hero in 2022

Reprinted in 2023

© Allary Éditions, 2019
Translation © Anne-Sophie Marie, 2022

Published by special arrangement with Allary Éditions in conjunction
with their duly appointed agent 2 Seas Literary Agency

The right of the above author and translator to be identified as the author
and translator of this work has been asserted in accordance with the
Copyright, Designs and Patents Act 1988. British Library Cataloguing
in Publication Data available.

Printed and bound by Imprint Press

ISBN (PRINT): 978-1-91505-470-8
ISBN (EBOOK): 978-1-91505-471-5

CONTENTS

Our perception is biased, our attention span is limited, our memory is deceitful. And yet, we have a coherent "world view". We can thank our brain for this, as it performs "tricks", mechanisms that enable us to understand the multifaceted, complex world we live in, and to share it with one another.

The brain, which shelters our knowledge, operates through estimates. The outcome is that our knowledge of things and of the world is always relative. The brain creates templates for absolutely everything: our friendships, our romantic relationships, our concept of work, our political opinions... Often unbeknownst to us, the brain tells us stories that help us better navigate through the world. It can completely recreate childhood memories or prepare us for a potential danger to save our skin if this danger proves to be real; it makes us understand that a pile of wax in front of us is actually a melted candle... but it can all the same fool us with an optical illusion or a magic trick, make us fall into the *fake news* trap, or into "knowledge delusion". In this journey to the centre of the brain, we will study the mechanisms and methods of this mysterious and extraordinary organ, to discover when, why and how it plays tricks on us and on itself.

YOUR BRAIN IS PLAYING TRICKS ON YOU

FOREWORD

Cognitive science is a fairly recent field, currently in full expansion. Some degree of trial and error is therefore inevitable, especially when we take an interest in an organ as complex as the human brain. Throughout this book, we will proceed according to a principle we learnt from Isaac Asimov: the relativity of wrong. Contrary to popular belief, right and wrong are rarely ever absolute, but often rather relative. This is why we will give you the theoretical models that are currently the most reliable, in order to become better acquainted with your brain and to better understand your own self.

PART I

HOW DO WE SEE
THE WORLD?

1

DO WE REALLY, LITERALLY SEE THE WORLD WITH OUR EYES?

"Like all great travellers, I have seen more than I remember, and remember more than I have seen."
Benjamin Disraeli, British statesman

We tend to think that we see the world with our eyes and hear it with our ears, which is normal: our perception goes through our senses first. Yet it is first and foremost with our brain that we perceive the world.

The five senses and the brain obviously work together so that human beings can indeed perceive the world. But our eyes, our ears, our tongue and our skin are actually *receptors* which will transform signals reflected by the outside world (optical, acoustic, olfactory...) into electrical signals. It is these thousands of electrical signals that our brain will process and filter, and which will enable us to mentally reconstruct the world.

The human brain and the world's ambiguities

Let us analyse an experience each and every one of us has had: an optical illusion. This term is deceitful, because it leads us to think that our eyes are the ones deceiving us. Yet the victim of the illusion is often our brain.

Look at this image:

Without thinking, does the black figure seem to be facing us, or does it have its back to us? Are you above it, or below? You're hesitating...

Now look at the image below: the individual clearly seems to be facing us, their elbows leant on the barrier, and they're located above you. And now that you have this image in mind, look at the first version of the image again. The interpretation you make of it will copy the scenario that image (a) led you to see, and now the black figure appears to be facing you at a low-angle shot!

Now let's go to image (b). Look at it for a few seconds, as you have done for image (a). Then come back to the original image.

The black figure from the initial image now has its back to you, and you observe it from above.

Here are now all three, placed next to each other:

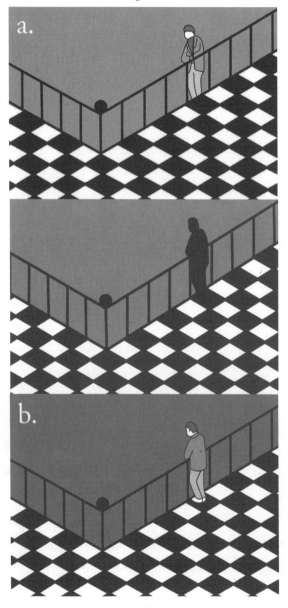

By looking at the top or bottom version for a few seconds, you can modify your perception of the central image as much you like.

How do we perceive the world?

Finally, only focus on the original version: now that you know the two versions stemming from it, you can easily change your mental perspective and see the character with their front, then their back to you, from high up, then from low down, without having to watch versions (a) and (b) of the image.

Let us now get into the specificities of this illusion in order to really understand how this image affects the human brain: images (a) and (b) are *stable* versions of the original image. There is only one way to interpret them. The original image, however, is ambiguous because it carries several ways of being seen – two, to be precise. The central image is therefore a bistable image.

When facing it, our brain doesn't possess enough information to solve its ambiguity and interpret it one single way. If, however, you stare at one of the two stable versions of the original image – that is to say, image (a) or image (b) – your brain will create a visual preconception and when you look at the bistable image again, you'll reduce its ambiguity and you'll either see a person's front (preconception (a)) or their back (preconception (b)) when looking at the black figure.

The brain has a need to interpret the signals the world sends its way in order to create a coherent and stable representation of the latter. This is called reduction of ambiguity: as soon as it's denied stability by being presented with ambiguous images (bistable or multistable), the brain proceeds to choose among the various options that reality contains.

This time, imagine that you're watching the first image (i.e., the bistable image) with a friend. Neither one of you has seen the stable versions of this image. Each one of you will reduce the ambiguity at hand in their own way: the figure seems to have their back to you, while your friend sees it as facing forward. You're actually both watching the same image, but you're seeing two different things. If you discuss this together, you won't come to an agreement, because your perceptions aren't the same, but each one of you is profoundly convinced they're seeing the image as it is. You're even unable to see what the other sees.

A bistable illusion was the talk of the town on social media back in 2015, brilliantly asking if we really share the same world. A Tumblr user named "Swiked" posted the photo of a dress with lace insets, followed by this comment: "Guys please help me – is this dress white and gold, or blue and black? Me and my friends can't agree and we are freaking the fuck out." Following this, the image went viral, and the entire world, divided, debated the colour of the dress for several days! If at the time you took part in this debate, you would have likely thought that the half of the world who didn't see the dress the same colour as you did was *wrong*. But now you understand that neither one of the two groups was right or wrong: there were simply two ways for the human brain to reduce the ambiguity.

What we learn from these two examples of bistable illusion is that human beings tend to blindly trust their perception, to the point of considering it to be shared by everyone.

When it filters, processes and interprets the stimuli that the world sends back, the brain constructs a generalized vision of the world, ceaselessly making, without noticing it, assumptions on the way the latter works. It constantly works to reduce ambiguity – and not only in cases of bistable illusions – in order to present us with a stable and coherent reality.

There's a blind spot in our field of vision, which corresponds to the place by which the optical nerve exits the retina towards the brain. We could assume there should be a "hole" in our field of vision where the light isn't received by the retina. However, in everyday life, our field of vision is whole because we have two eyes. But if we were one-eyed, or if we simply closed one eye, this would be altogether different.

So, close your left eye and watch the cross on the image below with your right eye, keeping your face at the centre of the page throughout. Gradually bring the page closer to your face.

All of a sudden – when the page is about 25 cm away from your eye – the black dot to the right of the cross disappears. This is due to the fact that it is situated on the exact blind spot of your retina, and your brain will reckon that the entire page is blank. And thus it will misrepresent reality.

Now do the same experiment with this image:

As soon as the black dot falls on your blind spot, the grey bar seems continuous. Your brain sees grey before and after the point: it fills the void with the same thing.

What magic tricks teach us

Magic tricks fascinate us. If they are universal, it's because they play with our brain's mechanisms, especially the one we've just shed light on: ambiguity reduction.

This is the case, for example, with the coin trick. A magician catches a coin between the thumb and index fingers of their right hand, then slowly places it in their left palm before closing their fist, which they then direct towards you, asking you to blow on it.

They open their hand in a theatrical way: the coin has disappeared, like magic! And they don't stop at that: they will strive to make the coin reappear behind your ear or in your pocket.

In reality, the coin was never placed in their left hand. The magician performed what we call a palming: they behave as though they were placing the coin in their left hand, while they have kept it in their right palm. All this is done very carefully, because the magician isn't trying to deceive our eyes, but rather our brain and its logical interpretation of objects' movements. Human beings rely on the brain's perception of the world: they think they have seen the coin move from one hand to the other, so they won't understand how it can end up behind their ear.

There is therefore a breach of coherence: something unreal has just happened. This is what we call "magic".

Our brain, from the moment we are awake, spends its time making assumptions on reality, interpreting it, filling its voids. It does it from our youngest age, unbeknownst to us. The table where we eat, from whichever angle we look at it, whatever light there is in the room, still remains the same table. Similarly, if we place an object in a specific location, we know that it will not move from there. It's the object permanence principle. It's thanks to this constant interpretation and reconstruction of reality (unsurprisingly incomplete), that reality itself seems so real, that objects seem to be fixed and unchanging things. This is why we are fooled by the coin trick.

Curious to find out through which psychological mechanisms their tricks managed to fool people, some illusionists have collaborated with neuroscientists. Teller, one of the greatest magicians of our time, has for example contributed to an article in *Nature*[1] in which existing links between magic and human beings' sense of the world are revealed. Teller begins with a famous magic trick: the cups and balls. The spectator faces three cups and balls which the magician makes disappear or move from one cup to the next "like magic".

Teller recounts that one day, just before going up on stage, he realized that he'd forgotten his cups and balls at home, and so found himself having to use what he had in his dressing room: transparent cups, and balls he shaped out of paper tissue. While he feared that the public would discover the inner workings of the trick, the spectators were even more impressed than usually. "The eye could see the moves, but the mind could not comprehend them," he said in an interview for the magazine *Wired*.[2]

There's a famous expression: "We don't see the world as it is, but rather as *we* are." It is a profound truth that the work of cognitive science confirms today: the world constantly reflects a multitude of signals, and we reduce their ambiguity by choosing what we want to see. Thus, little by little, our interpretation of the world shapes us psychologically, culturally and socially.

This doesn't mean, however, that we can see what we want to see all the time; in other words, that nothing really exists and that we are free to shape our own reality simply by imagining it in our head: in the context of the optical illusion we've previously examined, I'm free to see a figure facing forward or with their back to me, but I can't see this figure as a tree or a banana, for example. Reality exists and it's intangible, even when we may not know how to understand it without our brain interpreting it first.

Filling the void

The ambiguity of the signals received always puts us in an uncomfortable position of uncertainty. Thus, if our perception is missing an element to get rid of an ambiguity, our brain is going to want to fill that void. Descartes writes in the second of his *Meditations*: "What do I see from this window, if not hats and coats, which could be covering spectres or automatons merely moving thanks to springs? But I judge that these are real humans, through the sole ability to judge which is in my mind, what I believed I was seeing with my eyes." The eye doesn't see human beings under those capes and hats, but the brain restores them. Descartes understood that our brain "filled in the blanks" long before the first findings in cognitive science.

Here is another, amusing example of a "void" filled by the brain. How do you read this sentence: "Th15 is how you re rdeaing th15 lin3 R1gth now"?

You've probably read: "This is how you're reading this line right now." Your brain has just "recreated" meaning, while really this sentence means nothing. Your brain has put an apparent disorder back in order and has therefore chosen to prioritize its interpretation rather than sticking to the strict reality of what was written. A fine example of our brain's work: it prefers to give meaning to a block of letters and, by extension, give sense to things and to the world, rather than to linger in vagueness.

* * *

Our brain, which filters the myriad of ambiguous information that reality permanently feeds us, construes the world and recreates reality, often unbeknownst to us. In most cases, it's very useful and even vital. But this can also lead to mistakes that may be harmful to us.

Therefore, we need to shed light on the way our brain proceeds when it plays tricks on us.

2

HOW THE BRAIN TELLS US STORIES.

"It's like everyone tells a story about themselves inside their own head. Always. All the time. That story makes you what you are. We build ourselves out of that story."
Patrick Rothfuss, *The Name of the Wind*

When blind people think they can see

The brain, by recreating the world, enables us to find coherence within us and with our environment. To achieve this, it sometimes needs to invent things. In the case of some neurological diseases, this ability to invent is pushed to the extreme: we call it confabulation.

Anton Syndrome, also called visual agnosia, reveals exactly how far the brain can go down the path of confabulation. Only twenty-eight cases have been recorded to date, but they are noteworthy. Visual agnosia is a neurological awareness disorder which affects the patient's vision. Their blindness is located at the cortical level and not the retinal one: the retina absorbs light, but the brain isn't able to transform these stimuli into images. The patient is "brain blind" but they are completed convinced they see properly.

A case was reported in 2007:[3] a six-year-old child who had lost his ability to read and would miss objects when he tried to catch them and often fall down. His parents put him through a visual acuity test: the child was incapable of reading the largest letters from one metre away. He obtained a result under 20/2,000: he was completely blind… while claiming his sight was perfectly accurate. When asked why he had bumped against walls or couldn't grab objects within his reach, he invented justifications afterwards along the lines of "no, I didn't hit myself" or "it was a game".

It's important to make it clear that someone with visual agnosia does not *lie*, because lying is an intentional process. Their brain tells them stories, to the extent of making them truly think that their sight is normal.

The brain: singer-songwriter

The brain is made of two hemispheres (left and right), linked by a structure called corpus callosum (or callosal commissure).

Corpus callosotomy was performed on epileptic patients until recently. It is a surgical procedure which consists in partially or fully sectioning the corpus callosum in order to disconnect the left hemisphere from the right hemisphere. This practice was first developed in the 1950s following the work of neuropsychologist and neurophysiologist Roger Sperry, who had discovered, while sectioning the corpus callosum of a monkey, that this had practically no significant impact on its general behaviour.

Contrary to popular belief, we do not have a creative right brain and an analytical left brain, nor do we have an artistic right brain and a mathematical left brain.

Some functions are nonetheless lateralized, in other words, lodged in one of the two hemispheres, but most are bilateral and therefore present in both hemispheres. It is for this reason that, with human beings as it is with monkeys, callosotomy doesn't really affect brain function.

Language is one of those aforementioned lateralized brain functions and is often located in the left hemisphere (there are "righties" and "lefties" when it comes to language lateralization in the brain). Michael Gazzaniga, who was working on these questions with Sperry, wanted to know if it was possible to only communicate with one half of the brain, without the other half being aware of it. To understand the experiment, it is necessary to know that information received by our left eye is processed in the right hemisphere of the brain, and vice-versa.

Gazzaniga asked two callosotomized patients to cover their left eye (the eye that didn't have access to the "language" function) and only to look at an image with their right eye, linked to the left hemisphere of the brain. Afterwards he asked them to say what they saw, which they did with no difficulty. Gazzaniga then showed them a new image, and asked them to look at it with their left eye (linked to the right hemisphere of the brain, deprived of language): the patients could not express what they saw. Gazzaniga asked them to draw the image: while they hadn't been able to verbalize what they had seen, they were capable of making a drawing of it!

Gazzaniga went further by developing a new experimental protocol. As with the previous experiment, he showed images on a screen to a callosotomized patient who had covered his left eye. After having shown him a first image representing a chicken's foot, Gazzaniga asked the patient to choose which, among several photos, corresponded best to what he'd just seen. The patient pointed to the photo of a chicken with his finger. Gazzaniga asked him why he'd chosen the chicken, to which he instantly replied: "Because the image represented the foot of a chicken." The experiment was reiterated, this time with the left eye (i.e. the eye which doesn't have access to the "language" function). The screen no longer showcased a chicken's foot, but a house covered in snow. Among the photos he was later presented, the patient rightfully singled out a snow shovel. But when he had to explain his choice, things became complicated. He didn't reply: "Because I've just seen a house covered in snow." Nor did he say that he didn't know. He said: "Erm, well... the shovel's to clean the chicken coop!"

Here is an example of extreme confabulation: the callosotomized patient goes as far as making up a reason to justify his choice after the fact and using elements within his reach, in order to keep his own coherence.

To reduce the ambiguity of the world and to create a stable and coherent version of it, the human brain interprets reality and constructs narratives which are at times preposterous, to the point of inventing them if this is needed. It plays an active role in the *present* perception of the world, which it never ceases to recreate.

What if this capacity to confabulate extended to our memories as well?

Rewriting the past

Memory plays a major role in the elaboration and shaping of our emotions, our beliefs and our convictions. It doesn't work like a photo or a video camera which would merely record and objectively archive our memories: it actually recreates them.

Try to picture the last time you took public transport. The first thing to point out is that our brain hasn't been able to store all the information regarding the other travellers: their precise number, age, clothing... Yet you're not picturing an empty bus or train, nor shapeless and faceless phantom passengers. Try to remember the distinctive traits of these travellers more precisely. Unless a specific detail caught your attention at that very moment, most of the people and clothes you're now seeing in your mind's eye have been entirely recreated by your brain, which, in order to do this, resorted to what it believes are the standard physical features or clothing styles of the average traveller. Always eager for a coherent and stable world, our brain completely fabricates recollections, which gives our memory the texture of reality. So how much can our memory be swayed, and when can this malleability prove detrimental?

Until recently, in the United States the eyewitness of a crime could alone tip the outcome of a trial one way or the other, by presenting their recollection of the facts as the sole "piece of evidence". Innocent people have been imprisoned or sentenced to death for crimes they had never committed because of a witness's faulty memory. The works of Elizabeth Loftus, one of the most distinguished experts on human memory research, shook the foundations of eyewitness testimony to the ground and changed American courtroom procedures.

Elizabeth Loftus sought to find out how reliable our memories were and if it was possible to manipulate them, or at least to intentionally direct them. In 1974, she ran an experiment on the memory's reconstruction of an event[4] with John C. Palmer. Loftus and Palmer showed the video of a road accident to a panel of 150 students.

A week later, they called them in again and asked them if the car windows had been broken during the collision. The students were split into two groups. Asking the question to the first group, the two researchers used the word "smash": the car was *smashed* against the wall. The word planted the seed of a violent collision. With the second group, they used the word "*hit*": the car *hit* the wall, denoting a lower impact. In the video watched by the students, we can clearly see that none of the windows were broken during the accident. And yet the vast majority of the first group asserted that the windows had been broken at the time of the collision. As for the second group, it was the other way around.

By changing a single word in their question, Loftus and Palmer successfully modified the participants' recollection of the accident. Following this experiment, Loftus wanted to define what she calls the misinformation effect: precision and reliability lost by a subject due to information received a posteriori.

In some countries, such as the US, when a person is the victim of an attack, the police often uses a panel of individuals who look alike, among whom the victim has to identify their attacker. Loftus noticed that the victims almost always chose someone... even in situations where the culprit wasn't part of the panel: their recollection has been altered by the implicit suggestion that the culprit is necessarily among the men or women presented to them!

Following the publication of Elizabeth Loftus's research, lawyers and legal experts launched the "Innocence Project" in New York in 1992. For the past thirty years, Innocence Project has brought the cancellation of close to 75 per cent of verdicts given after a line-up or because of an eyewitness, using DNA tests as evidence of the innocence of the people condemned. This is how Kirk Odom was found innocent after twenty-two years behind bars: he had had the misfortune of being identified by one of the victims as guilty of a kidnapping and rape.

Elizabeth Loftus also wondered if there was a way to plant false recollections inside our memory. She looked into repressed childhood memories that resurface in adulthood, often during therapy or psychoanalysis sessions.

Elizabeth Loftus gathered a panel of men and women who had never experienced any trauma linked to abandonment during childhood.[6] Using suggestion techniques equivalent to the ones from the car accident, she successfully convinced 25 per cent of them that they had got lost inside a shopping mall when they were little. In many of these cases, the participants even invented details throughout their narrative, embellishing the traumatic moment that had never taken place.

These suggestions techniques are sometimes used maliciously. In 2017, Marie-Catherine Phanekham, a forty-four-year-old physiotherapist, was condemned to one year in prison and a twenty-thousand-euro fine for having extorted considerable sums from her female patients after planting false memories of childhood rape, incest or violence.[7] The aim of the procedure was to invite them to overcome their trauma – which by then had become real, even if the memories were false – through a long and costly therapy. Mrs Phanekham used suggestion techniques to take advantage of her female patients for her own material gain.

Gaslighting[8] is another form of cognitive hijacking which leans on memory manipulation: it consists in making a victim doubt their memory or their actual mental health by presenting some facts in a truncated manner, editing a few elements of the original memory, telling them they've made it all up, or that they're losing their mind. This form of emotional abuse can take many shapes. For example, in the professional arena, when a boss heavily criticizes an employee in order to push them over the edge, as a form of psychological harassment, but refers back to the event a few days later as banter, telling that same employee they shouldn't take it badly, the boss confounds their employee in order to make them doubt their own memory of that event. This employee will then continue to fear their boss, but will never dare to file a harassment complaint.

It was also the case with members of the "Ligue du LOL" (League of the LOL), a largely male French Facebook group created in 2010, when they decided to harass female journalists and feminist activists online "for a laugh". This however had nothing to do with childlike fun and games: some members of the Ligue du LOL used pornographic images Photoshopped with the faces of female journalists while others harassed them with hateful messages. Today, their line of defence is simple: "It was banter, we were young." Lucille Bellan, a journalist, recalls the harassment she suffered in a *Slate* article: "It's difficult to come to terms with being a victim. Especially when everything can be covered up with the veil of humour. On bad days, you keep telling yourself: 'maybe I didn't understand this properly', 'my paper wasn't that good', and then you just take it."

These memory-manipulation techniques have also been used to political ends. This was for example the case of interrogations led by the Soviets during the Great Purge in the 1930s: the idea was to break prisoners accused of alleged deviationism and treason towards the party. In order to make them admit to imaginary crimes and keep them away from power, the prisoners were subjected to continuous and intense pressure through intimidation, bullying and psychological and physical torture (waking up in the middle of the night, constant electric lights in their cell, relentlessly repetitive questions asked in different ways etc.). They started to question the truthfulness of their own past, of their memories and, ultimately, of their innocence. Broken and exhausted, they ended up believing in imaginary crimes themselves which they admitted to have committed, leading them most often to the firing squad. Costa-Gavras dedicated a film, *The Confession*, to it. The film revolves around a former member of the international brigades, played by Yves Montand, who is interrogated and psychologically tortured at length. The Soviets end up convincing him he has betrayed the party during the Spanish Civil War, while in reality he had distinguished himself with his heroic deeds.

Memory can be a precious tool, especially for psychoanalysts or psychologists, but it's a tool one must handle with extreme caution because, even without intending to do so, it is possible to modify it retroactively.

We don't always remember the choices we make, but we justify them.

When we make a decision, we think that the mechanism triggered is the following: several options present themselves → I think about the situation → I settle on a decision that I will be able to coolly justify later. But does this really happen this way in our brain?

Petter Johansson[9] and his team have established the following experimental protocol: they show passersby photos of two women (a brunette and a blonde; they have their hair up, but it's still possible to make out their hair colour), and ask them to choose the one they find the prettiest. Unbeknownst to them, the researchers switch the two photos and then present the passers-by with the image they did not select. The passers-by are then asked to justify their choice – that is, the opposite choice to the one they have made. In total, 74 per cent of participants don't notice the subterfuge and defend their choice, tooth and nail, explaining for example that they have chosen this face rather than the other because of the person's smile, the shape of her chin etc. Johansson and his team will give a name to this capacity for post-hoc justification by our brain regarding a choice we haven't made: choice blindness.

The results of this experiment caused great controversy in the cognitive science field. Other researchers asked to know in which conditions the experiment had been realized, and if any external factors could have affected the results.

But choice blindness was introduced in several experimental protocols after this, and scientific literature is teeming with examples showing the extent of our brain's post-hoc justification capacities. In 2010, the American researcher Lars Hall led another experiment on choice blindness. In the supermarket of a small US town, Hall and his team set up a fake local produce stall. Dressed as salespeople, they presented two types of jam and two types of tea, asking the shoppers to share their opinion on the jam-tea combo they preferred. Let's highlight here that the two jams had two distinctive tastes, clearly different from one another – one was apple-cinnamon and the other citrus fruit – and that the pot had a trick: it could be opened from both ends, with each side containing a different flavour.

Once a shopper had tasted the various jams, drunk their cup of tea and made a choice, the researcher turned the pot over unbeknownst to the shopper and asked them to try their preferred jam again in order to justify their choice. Only a third of participants detected the change in taste. All the others justified their choice without realizing they were pleading the case of the jam they had not chosen! When the trick was revealed to them and the experiment explained, reactions varied from surprise to utter disbelief.

* * *

We are not exact beings: our brain often plays tricks on us and sometimes pushes us towards a mistake. If we do make mistakes, it's also because we build our world through approximations. But approximations aren't inherently bad: they are at the heart of our reasoning, of our ability to predict, and also at the heart of most of our thought and action reflexes.

3

WHY DO WE LIVE SO OFTEN BY APPROXIMATION?

¯_(ツ)_/¯ *

It's a Saturday night, you're with friends and the question of whether to order sushi or pizza comes up. In front of the general indecision, someone suggests a coin toss. If at that moment you were asked the probability of the coin falling on heads or tails, you'd rightly answer that it's fifty-fifty.

Now imagine, but this time in the context of a game, that you're being asked to perform one hundred coin tosses and note the obtained result each time, as well as the probability that the coin will fall on heads or tails on the next throw. You know that this probability is fixed and that it will always be fifty-fifty, independently of the number of throws.

The coin falls on tails the first time, then the second, and the third... It's surprising, but you consider it a

---

* This is not a mistake, but a symbol of approximation. Up to the reader to interpret it!

41

coincidence. You continue to think that the coin has a 50 per cent chance of falling onto one side and 50 per cent onto the other side on the next throw.

Now imagine that you're on your fiftieth throw, and until this point the coin has always fallen on tails. Is the probability of the coin falling on tails during the next throw still fifty-fifty? It is, because as we were saying earlier, the probability is fixed. But doesn't it seem more reasonable to realign our way of thinking to the reality imposing itself on us by suspecting for example that the coin toss has been rigged? You will find it hard to believe that you are the witness of a statistical miracle.

The brain's ability to establish mental patterns and fine-tune them little by little when in the presence of certain events bestows upon it an essential power of anticipation called inference. It is the ability to draw up a prediction on the future from our observations and our knowledge of the world, and to act consequently by choosing the most appropriate strategy for each situation.

Inference, or the art of finding a cab on New Year's Eve

Imagine you're looking for a cab. The probability of finding one straight away depends on several factors: the neighbourhood you're in, the time, the date, the weather, the traffic. If it's New Year's Eve or another festive occasion, the "everyone goes out that night" factor can become much more important than all the other factors mentioned above put together. Finding a cab becomes an almost impossible mission.

Nevertheless, if you've ever found yourself in this situation, you will be able to anticipate the taxi shortage and find another way of getting home: you're going to take the metro because you know it runs all night long, or you're going to ask your friends who drove over to give you a lift. You've updated your "finding a cab" template by adding information accessed from prior knowledge. You can then develop a reasoning that will enable you to have the highest chance of seeing your actions crowned with success.

Inference remains a form of approximation: it's impossible to predict the future. This capacity to anticipate, though it's not 100 per cent reliable, is nonetheless viable and necessary to us. Various sciences work according to this principle: meteorology consists in predicting what the weather will be like from the observation of satellite images and data from years past. In oncology, inference is used to evaluate a patient's risks of developing cancer in order to anticipate the onset of the disease. For example, from the age of forty, women are encouraged to have a mammogram every other year to anticipate an eventual breast cancer and be able to treat it as quickly as possible, because statistics show that women are more subject to breast cancer after that age.

The handshake

Each daily action requires immediate and unconscious decisions. Climbing up a staircase, applauding, hammering in a nail... all this implies decision-making (lifting the right leg then the left leg, bringing both hands closer to each other until they clap and repeating the process, raising the hand holding the hammer and hitting the nail we hold between the other hand's index finger and the thumb...). Let's study the case of a handshake in more detail, as it also involves you taking the action and movement of another person into consideration. Imagine that you're going to a job interview. The head of HR greets you by holding out their hand: you extend yours in response. You've automatically shaken their hand, without thinking about the exact angle of your arm or the pressure applied on their hand. Your brain is used to trigger the "handshake" action since it's already had to do it hundreds of times before. And yet it works each time. We call "heuristic" this reflex founded on a quick and approximate perception of reality, but which works more or less well.

Most of our everyday actions are heuristics, but thought has also its own: for example, our brain often rounds the time up to an easily memorable and com-municable number. If it's 8:27 p.m. and you're asked the time, you'll probably answer "half-past eight". Moreover, some preconceived knowledge of the world which triggers an intuitive action can also be called heuristics: when you see big dark clouds in the sky as you're about to go out, you will intuitively think that it might rain and you'll decide to arm yourself with an umbrella.

These heuristics enable us to prepare within the limits of our attention and of our mental faculties: we do not have the attention capacities, the time or the energy to assimilate all the information we receive in a given situation before making a decision.

Regarding the use of heuristics in decision-making, I would recommend an article published in 1974 penned by Amos Tversky and Daniel Kahneman, which earned the latter the Sveriges Riksbank Prize in Economic Sciences in Memory of Alfred Nobel.[10]

When intuitive thinking misleads us

Heuristics enable us to carry out those little actions we're not aware of and which serve us daily. But there are cases where too fast and approximative a thought reflex can lead us to making mistakes. These thought deviations leading to misjudgements as well as illogi-cal or irrational interpretations of a given situation have been named cognitive biases by psychologists Daniel Kahneman and Amos Tversky. This is how in some cases we make decisions too quickly by leaning on a limited number of elements which we consider representative of a situation. They call this representativeness bias.

In the context of an experiment, Kahneman and Tversky wanted to show that we occasionally prioritize "customizing" information over statistical information. They presented profiles of individual types, only describing a few specific traits of their respective personalities. Here is for example the description made of Steve: "Steve is very shy and withdrawn, invariably helpful but with very little interest in people or in the world of reality. A meek and tidy soul, he has a need for order and structure, and a passion for detail." After which Kahneman and Tversky asked their students to guess Steve's profession: farmer or librarian?

Given the personality stereotypes associated with these two professions, most of them answered that he must be a librarian. They forgot that on a global scale there are a lot more farmers than librarians, a factor which should have been taken into consideration in their reasoning and in their final choice. The participants used a heuristic method based on partial information (the personality trait) to get to a quick yet unspecific and potentially faulty answer, rather than adopting a more reflexive process.

Another common bias which can equally mislead us is the anchoring effect. In any given statement, we tend to retain the first piece of information given to us. If, for example, during a job interview, you're presented with a first candidate who's "nice, serious, but a bit prone to anger" and a second one who's "a bit prone to anger, but nice and very serious", you'll tend to have a positive impression of the first one, when the two candidates have the exact same character traits.

Following the works of Kahneman and Tversky, hundreds of cognitive biases[11] have been identified and researchers continue to list new ones regularly. Two of the biases we speak about the most in our age of misinformation (new term used to talk about fake information supported by no facts) are the confirmation bias and the anecdotal evidence bias. Confirmation bias pushes us to only take into consideration information that reinforces our opinions, our convictions and our beliefs, and to reject as false all other ideas which could be presented to us. Anecdotal evidence bias occurs when we use an anecdotal example to justify our reasoning. It is what people do when they want to ban certain video games under various pretexts. For example, the violence they're meant to generate. To that end, they cite isolated cases of young people who have committed acts of violence and who play those games. Yet, by doing this, they ignore all the other teenagers who also play but have never committed violent acts. This is evidence through anecdote. We'll come back to it in the second part, the important thing for the moment being to understand what a cognitive bias is and how it takes shape in our minds without us even becoming aware of it.

Intuition vs reflection: can we only think in these two ways?

Following his theorization of the cognitive biases misleading human beings, Kahneman offered a theoretical model for the human thought process. His hypothesis is that we have two thought systems: system 1 is heuristic, intuitive and fast but subject to errors, and system 2 is reflexive and logical, and therefore slow and requiring more effort, but more reliable. For example, when I want to know how much 2 + 2 adds up to, I launch my system 1, while I use my system 2 to resolve 108 x 82.

It is sometimes necessary to go from system 1 to system 2: let's imagine a croissant and a sweet cost 1 pound and 10 pennies. The croissant costs 1 pound more than the sweet. How much does the sweet cost? According to system 1, you will start by automatically thinking: "10 pence". However, if you activate your system 2 and you take a sheet of paper and a pencil to note down the process, you'll see that your initial and intuitive answer is wrong. The croissant actually costs £1.05 and the sweet £0.05.

The virtues of intuition

For Kahneman and Tversky, the reflective model, more reliable and less prone to error, is superior to the intuitive model. But Gerd Gigerenzer, one of their most serious detractors, stipulates[12] that, contrary to preconceived ideas, an intuitive and thus biased system can beat a lengthier reasoning system. He explains that in certain situations, an overly lengthy reflection, an overly exhaustive analysis of situations can hold us back and inhibit our choice: the more information we have and the more options present themselves to us, the more difficult the choice.

For Gigerenzer, the brain is an "adaptive toolbox". The researcher presents the following situation:[13] you're the head of a big company which possesses a portfolio of 100,000 clients. You would like to carry out an advertising campaign targeting "passive" clients (those who don't buy your products often enough). To target the passive clients, you can choose a complex solution and use an exhaustive statistical model. You'll obtain a value for each client which will indicate the likelihood of them purchasing your products in the future. Or you can use a heuristic solution which will probably be less precise but quicker: the hiatus heuristic. From your portfolio, you select a client who hasn't purchased anything for nine months, who you thus consider passive. Now all you have to do is select all those who haven't been part of your clientele for even longer than him: those are the ones you will target first. In that case, the heuristic method is quicker, but no less efficient than the statistical method, since you will target a panel of clients matching your concept of an inactive client.

Kahneman's hypothesis[14] has been nuanced by recent research showing that the brain functions in less of a binary manner than this theory would have us think.[15] The main charge against Kahneman's model is that it rests its case on a logical error* called a binary reasoning error. Human beings have a very strong penchant for everything that works in pairs: good and evil, left and right, hot and cold. Dualism is a mechanism to reduce ambiguity, which is applied to concepts too complex to comprehend other than reducing them to two definite and antagonistic realities. We will thus think that the human brain also functions according to this model: intuitive/reflective.

---

* Another form of error comes to regularly influence our reasoning: what we call logical errors. We call Sophism a reasoning that seems logical but which is fallacious. For example, Mark is violent and he has black hair. David is violent and he has black hair. Therefore, people who have black hair are violent.

Yet there is practically no empirical evidence to support the idea of a binary functioning of the human brain.

Some research[16] reinforces Gigerenzer's theory: in a given situation, "slow thinking" and therefore adopting a type 2 reasoning, seemingly more reliable and accurate, can reduce the relevance of a choice, or the satisfaction stemming from it. You're at the restaurant and the person who's with you is taking a long time to choose their dish, changing their mind several times before being obliged to order. You know from the start what you want to eat. You're likely to take more pleasure from this meal than your friend who won't stop wondering if they should have picked the sole rather than the hake.

Moreover, adopting a type 2 reasoning doesn't always prevent mistakes. Some cognitive biases only affect system 2: you play a game of chance and you're winning. You think and you deduce that you're on a roll and that luck will therefore continue to be on your side. But nothing is less certain! You're being subjected to a form of optimism bias.

The human brain functions in a dimensional way, rather than a binary one. It would be less similar to a switch which only has *on* and *off* functions, than to the volume dial of a radio which can be gradually regulated.[17]

Kahneman's model can be useful to us if we keep in mind that it's an abstraction used to simplify our comprehension of cognitive functioning: it remains "less false" than the models we had before. We simply can't forget it's not 100 per cent reliable.

The imprecisions of this model brought about new questions which enabled us to sharpen and disprove some of its aspects and move forward in our quest to reduce our error margin. That's because science functions according to a correcting principle. It's by explaining the mechanisms of the scientific method and by highlighting the merits of failure that we can improve our mental flexibility and no longer fall into the traps our environment sets for our brain every day.

* * *

We must therefore avoid considering heuristic operations, and the cognitive biases they expose us to, as entirely negative. Nowadays the scientific press takes a keen interest in the cognitive biases that are detrimental to rational thinking: the confirmation and belief biases mentioned above, but also negativity bias, self-serving bias, knowledge illusion and learned helplessness, which we will talk about in Part II.

Consequently we often reduce a human being's behaviour to a list of mistakes that need fixing: proof being that, even on the state scale, or the World Bank scale, we advocate debiasing.[18] However, as we've seen, cognitive biases aren't immutable, they're not absolutely positive or negative, but dependent on several factors: they're *contextual*. It would be more interesting to learn *why* some biases emerge in certain situations, rather than fighting them at all costs.

PART II

MY BRAIN, THE OTHER BRAINS AND THE WORLD

4

STRESS: YOUR BEST ENEMY

"We live well enough to have the luxury to get ourselves sick with purely social, psychological stress."
Robert Sapolsky, American scientist

Nowadays, stress is considered to be the scourge of the century: in 2013, according to a study financed by the European Union, costs related to stress in Europe were estimated at 617 billion euros, and stress at work affected 25 per cent of salaried employees in the EU, which is huge – and this percentage has continued to grow ever since.[19]

Prolonged exposure to stress has many harmful effects on our physical and mental health: it generates anxiety, can lead to depression, causes sleep issues, back aches and digestive problems, weakens our immune system and causes ulcers.

Stress also has an impact on our psychology: it modulates our ambiguity reduction and our coherence, and leads to the emergence of numerous negative biases: when we're stressed, we feel our brain working against us.

Stress often has paradoxical consequences on human beings. For example, why can't a high-school student manage to fall asleep the night before an exam, when a good night's sleep would be a lot more useful? In the same way, why will we find ourselves suddenly drawing a blank in the middle of a presentation, in front of an entire group of people, when we were properly focused on our speech up to this point? Why do we suffer from migraines when an important decision has to be made? Why do we lose our appetite when we learn stressful news? In other words: why does our brain turn our body inside out when a stress reaction is triggered?

However, stress is at the outset a positive mechanism that our brain activates in order to help us survive. We are going to see how and why it has transformed into a handicap for human beings today.

Imagine you are in the savanna, discreetly following a zebra.* The animal is peacefully living its zebra life: it grazes and it walks around. Its stress level is at its lowest. Suddenly, the zebra notices a starving lioness from afar: its stress level skyrockets immediately.

American researcher Karl Pribram ventured the hypothesis that animals possess four instincts that enable them to survive.[20] He called them the 4 Fs: Feeding, Fucking, Fighting and Fleeing. What interests us is the fight-or-flight sequence, because

* A homage to Robert Sapolsky and his book *Why Zebras Don't Get Ulcers*.

58

it corresponds to the first response to stress among vertebrates.

In the case of the zebra, the stress peak (fight or flight) is triggered at the first sight of the lioness. The zebra's sympathetic nervous system automatically reacts by increasing its cardiac and respiratory activity as well as muscular tension: it needs this to flee, or if it has to fight. Conversely, it inhibits its digestion, immune system and libido, all useless to its immediate survival. All of the zebra's resources are mobilized towards a single goal: to survive. If it had remained relaxed when spotting the lioness, the zebra probably wouldn't have survived very long. With the lioness, the stress peak is equivalent to the zebra's, but hers is triggered by a fear of starving to death: when she sees the zebra, the lioness mobilizes all her energy to try to catch it and eat it.

Stress is therefore an essential function for all vertebrates because it privileges survival: if you're in danger of death, it's useless to invest energy into digestion or libido, or in combatting a virus. On the other hand, it's necessary to have the most operational muscular system one can possibly have. It's important to note that, with animals, the stress peak doesn't last long: once the danger is gone, their parasympathetic nervous system (the inhibitive complement of the sympathetic system) takes over and everything returns to normal.

Let us now transpose the zebra scenario to our ancestor, the *Homo sapiens sapiens*, before their adoption of sedentary life, because they had the same biological system and the same cerebral capacities as we have. Let's imagine one of them peacefully picking wild berries. Suddenly, they hear a noise coming from the foliage. This noise is an ambiguous, bistable stimulus: two options are offered. Either it's a predator approaching, or the wind blowing through the leaves. The brain of the *sapiens sapiens* can opt for the "predator" option. It then triggers a stress reaction: the nervous system is mobilized towards immediate survival, muscles contract and breathing accelerates, they abandon their picking and dart as fast as possible for a chance to survive. If it was only the wind in the end, too bad: they will have run for nothing, lost their berries and will be dripping in sweat.

If they choose the second option and continue their picking thinking the noise comes from the branches, the *sapiens sapiens* is exposed to the risk of being devoured by a famished predator. Of these two options, the first presents an obvious evolutionary advantage compared to the second one: *sapiens sapiens* prefers to overreact over the risk of being devoured. Those who survived and whom we can call our ancestors are the *Homo sapiens sapiens* who chose to react to false alarms rather than to ignore them, and who opted for the risk of triggering a stress reaction for nothing rather than exposure to the danger of being eaten.

Now imagine our ancestor *sapiens sapiens* getting ready to sleep in a cave. Just before going to bed, they see the shiny eyes of a lynx from afar... They will not fall asleep peacefully. On the contrary, they will remain vigilant. In fact, they will spend all night in half-slumber and the slightest noise will awaken them to enable them to check if the lynx might have come closer.

In the same situation, if they're not alone in the cave and they don't see the lynx themselves, but read stress on the face of one of their peers, they will probably become stressed as well, through anticipation, in order to be able to start running if this becomes necessary: the contagiousness of stress is a very useful social signal in any danger context.

For close to 300,000 years, human beings have thus been conditioned to have a quasi-automatic stress reaction as soon as they are in an ambiguous situation in which their survival is at stake.

We can therefore define stress as a physiological reaction facing a physical danger which, for millennia, has enabled us to move about and protect ourselves as best we could in a hostile environment, to survive and to evolve. However, at a certain point in human evolution, human beings became sedentary, going from a lifestyle where danger was for the most part linked to predators all around them to a modern lifestyle where dangers have become less immediate and less physical (deadlines, poverty, workload, public speaking...). Today, the dangers we're confronted to tend to be more of a psychological nature (although unfortunately for some of us more than others, such as women or members of minority groups, the threat of physical danger is still all too real): if you're asked about your main source of stress, you're unlikely to respond it's a starving lynx, but rather that it's bills piling up, taxes, your higher-up at work, etc.

On the evolutionary scale, this lifestyle change happened relatively quickly, and this stress hasn't really had time to adapt to the evolution of the kinds of danger human beings have been and are being exposed to. Now that physical dangers have been replaced by psychological ones, human beings, when they become stressed, use a tool no longer adapted to their needs.

Imagine the following situation: in the context of a work meeting, you've decided to announce to your boss that you have a groundbreaking idea which could boost your career. But if you're stressed at the moment you're speaking, your brain will trigger a fight-or-flight mechanism: your heartbeat will accelerate, your muscles tense up... though you neither want to flee nor fight your boss: all you want is to succeed with your presentation. But for your brain this is the equivalent of a large feline lurking around you. And since all your resources are going to be mobilized in one go towards your immediate survival, what was a priority for your brain only a few minutes ago (i.e. the content of your speech) is no more, and you may become the victim of a memory lapse.

If with animals or *sapiens sapiens*, the fight-or-flight reaction upon spotting a predator was a matter of seconds, with us stress can settle in for the long-term. Yet our bodies aren't built for it. Imagine your brain has to defend itself against the imminent attack of a predator for months, with very little respite: you'll exhaust yourself waiting. We are built to withstand intense but brief stress peaks. As the human body cannot bear to stay constantly alert (i.e. in a state of chronic stress), it eventually gives in and ends up in a state of burnout. Seeing the impact of stress on our vital functions and on our sympathetic system, we understand better why someone with burnout can end up bedridden and unable to move.

When someone is stressed out, it's useless telling them to calm down, because their brain is warning them that their life is in danger. This would be the equivalent of saying: "Come on now, stop wanting to survive, OK?" There are, however, techniques to fight against stress's harmful side effects. Meditation, yoga, stretching, coherent breathing techniques and sports are effective: by breathing slowly, you defuse the primal fight-or-flight reaction. Your muscles relax, your heart-beat regains its normal pace, you are mechanically less stressed. Relaxing your body is going to allow the brain to reduce the ambiguity of the stressful situation and make it less anxiety-provoking. In a stress-ful situation, the information communicated to the brain is: "this meeting must be extremely important, otherwise why would I have put my body on alert?" If consciously you "relax" your body, your brain will be more likely to say: "If my body is that relaxed, it's that there isn't really any danger." These techniques are mechanical and have nothing to do with any ener-gies or mysticism, as they directly reduce your stress levels by defusing your fight-or-flight reaction and, through this, your anxiety.

Stress and anxiety: same thing?

We tend to use the words *stress* and *anxiety* inter-changeably, while these two terms refer to two slightly different phenomena. Stress has an identified cause and subsides once this cause disappears. For exam-ple, if you're stressed due to an exam, you will be able to unwind once the exam is behind you. Anxiety, however, doesn't need a cause, and can therefore last indefinitely. In some cases, anxiety doesn't even have a definite cause to begin with: we can be anxious and not really know why.

Stress and anxiety influence our way of reducing the ambiguity of the world around us, as well as, therefore, our biases. Studies have shown that people with anxiety[21] reduce the ambiguity of certain words in a more negative manner than others. If we ask someone with anxiety what the word "mug" means, they will have a tendency to answer it means to attack (as opposed to the more neutral synonym for "cup" or slang for "face"). This is what we call interpretation bias.

Attention biases are triggered in people who suffer from phobias, another type of anxiety disorder. Someone with arachnophobia will be able to detect the presence of a spider in their environment more quickly than average. Once the spider has been spotted, they won't stop glancing at it to make sure it hasn't moved: it's what we call an attraction-repulsion pattern linked to a strong feeling of hypervigilance.

Stress and anxiety disorders are often a cause of the degradation of our relationships with other people because when we have anxiety, we systematically reduce the ambiguity in a negative way. Let's take the example of social anxiety disorder. It is expressed as an excessive fear in various situations, ranging from public speaking to more ordinary things, such as making a complaint at a shop or ordering a jug of water at the restaurant. People who suffer from this have a negative interpretation bias[22] of these situations, which makes them see negative intentions behind these actions or words (when there aren't necessarily any) more than the average. They especially interpret the gaze of others in this light, and thus perceive a negative judgement quasi-systematically.

I led the following experiment with a research team: on a screen, we showed faces of men and women expressing several emotions successively – disgust, joy, surprise, sadness – to people suffering from social anxiety disorder. We then did the same thing with people you did not suffer from social anxiety disorder. Using an eye tracker,[23] we were able to observe the way in which these two types of people explored the faces presented to them. Results showed that people with social anxiety disorder, after a quick scan, went quickly to the eyes before lingering on the bottom of the face, then came back to the eyes frequently and furtively. Meanwhile, people without social anxiety disorder explored the faces following the direction of an inverted triangle: from the eyes to the mouth (which corresponds to a normal exploration). Another person's gaze triggers the same attraction-repulsion and hypervigilance in a social anxiety disorder sufferer as a spider does in someone with arachnophobia.

Social anxiety disorder is a form of anxiety through anticipation which contributes to the desocialization of the subject: if you're subject to social anxiety disorder and you're invited to have a drink with colleagues or to speak in public, you will anticipate the situation and reduce its ambiguity by telling yourself there will obviously be a form of danger. Several days before the event, you will trigger a fight-or-flight mechanism and on D-Day you will be incapable of attending the evening in question because you will judge it to be too dangerous. By avoiding it, you run the risk of aggravating your isolation even further as well as reinforcing your phobia. A truly vicious circle.

We can be, to varying degrees, subject to different forms of anxiety which have an impact on our inter-personal relations. Someone who's afraid of certain population groups – of certain ethnic groups, for example – will risk interpreting their actions accord-ing to negative interpretation biases. In the United States, researchers from Yale University noticed that, at school, black children were expelled from class more often than white children.[24] They met several primary school teachers and asked them why. The primary school teachers said that the skin colour-expulsion correlation was nothing but a pure coincidence and vigorously denied any racist prejudice.

The researchers later showed the teachers a video of schoolchildren in a classroom and asked them to spot "problematic" behaviour among the children. Using an eye tracker, the researchers noticed that the teachers spent more time observing black children – more associated in their mind with troublemakers – than white children. When the teachers discovered the results of this experiment, they all stated they had not been aware of having adopted a biased attitude towards black children until then.

Another team of researchers[25] showed photos of white men and black men to American volunteers of all origins and all ages. In the photos, some of the men were pointing their guns at the participants, while others weren't. The participants had a button in front of them and the instructions were as follows: they were asked to press the button each time they were presented with an armed man, and to do nothing when someone wasn't armed.

Results show that the participants pressed faster when they were shown an armed black man than when shown an equally armed white man. The danger perceived at the sight of a photo of a person that was black and armed activated the stress mechanism faster for all the participants, regardless of their own skin colour. Due to an implicit anxiety-provoking bias, conditioned by the skin colour of a person and not by the gun they hold in their hand, the ambiguity is reduced in a different way and proves the existence of a societal prejudice in the United States whereby black people would be more dangerous than white people.

* * *

Next time you negatively interpret a situation or you form a negative opinion of someone, it would be useful to ask yourself if you're under "pressure". Your jaws are tense, your heart beats quickly: these are signs that must alert you of your state of stress and enable you to take a discerning distance in relation to what you feel, in order to adjust your interpretation or your judgement.

These signs are often imperceptible because our stress mechanisms are activated too quickly. They are, however, easier to detect in situations whose anticipation itself provokes stress, which leaves more wiggle room to correct hasty interpretations: when I know that I am stressed by the prospect of an exam, I can prepare ahead of time by doing relaxation exercises which will enable me to be calmer on D-Day.

* * *

Some of our "primal" mechanisms, such as stress, haven't adapted to our lifestyle changes, at least not yet – it would incidentally be interesting to see in how many centuries or millennia the human species will modify its relation to stress. Until then, we can train ourselves to better interpret our reactions to stress, to anticipate them and outsmart them whenever possible because they condition our relations to others and to the world, and they influence our opinions and beliefs.

5

THE ILLUSION OF OUR CERTAINTIES

> "You see, there's something frightening in
> this world: everyone has their reasons."
> Jean Renoir, *The Rules of the Game*

Thinking like a detective or like a lawyer

Let's suppose you're sensitized to the dangers of GMOs, and that you think they're bad for our health. If we ask you to choose between genetically modified corn and organic corn, with both being at the same price, you'll spontaneously go towards the one that hasn't been subjected to genetic modifications. Imagine now that you're shown an article which explains that nothing today proves that GMOs are bad for our health; there is a big chance you will either skim-read it or not read it at all, rejecting in advance the ideas defended there because they are at polar opposites with yours.

And yet, to this day no convincing study has proven the harmfulness of GMOs to our health. Since the dawn of farming, human beings have endeavoured to cross-breed fruit and vegetable species and thus to genetically modify them. In a famous Giovanni Stanchi painting, we see a watermelon filled with seeds, and whose flesh, rather white, is divided into six distinctive slices. This watermelon from the seventeenth century is quite different from the watermelons we eat today, whose evolution is the result of lengthy human intervention. What poses a problem nowadays are the conditions in which multinationals such as Monsanto operate, and the attack on biodiversity which stems from these practices when they are excessively industrialized.

If you had taken the time to read the article, you would have opened your mind to a new perspective in the debate on GMOs, and you may have re-evaluated your own judgement. In other words, your mind would have gained in flexibility. But you let yourself be blinded by your primary belief. The idea isn't to take a side for or against GMOs here, but simply, through this example, to take an interest in motivated reasoning, a bias which pushes us to preferentially believe ideas which are aligned with those we already have, and to spot eventual pitfalls to better foil them. This example highlights that when we discuss a topic dear to our heart, we spontaneously function like a lawyer who has already decided their client was innocent and who will bring forward everything that supports their defence by adopting *motivated* reasoning through a preconceived idea. It may be necessary to learn to more often adopt the attitude of an examining magistrate or of a detective, who follows the clues step by step to reach a well-developed solution. In other words, to adopt deductive reasoning. The idea is not to systematically reject our beliefs outright, but to sometimes place them at a distance, for the time it takes to take into consideration arguments that nuance or negate them.

During the 2004 presidential election opposing George W. Bush to John Kerry, Drew Westen,[26] a professor of psychology and psychiatry at the University of Atlanta, wanted to show that adopting motivated reasoning could push us to believe more easily in the truthfulness of what is in line with our beliefs, and to resist what could come to contradict them, most particularly within a political context.

Western invited thirty people, all very much involved in the electoral campaign: fifteen card-carrying Democrats and fifteen card-carrying Republicans. The experiment was developed in three parts. Western first read a declaration by Bush on an important topic within the context of the campaign to the thirty people, such as, for example, the war in the Middle East. Then he quoted another Bush speech, in which he contradicted himself. He did the same for Kerry: a declaration on the environment followed by a contradictory statement. Finally, he quoted a "neutral" personality (an actor or a sportsperson) who had also contradicted themselves.

Then, Western asked the participants to rate the level of seriousness of these contradictions.

As the results show, the Democrats deemed Bush's contradictions a lot more serious than Kerry's, and vice versa. However, when this was a politically "neutral" person who contradicted themselves as much as both candidates, Democrats and Republicans equally deemed this contradiction to be hardly serious. Parallel to this, Drew Western recorded the cerebral activity of the participants through functional magnetic resonance imaging (visualization of cerebral activity via magnetic resonance). Results have shown that Democrats and Republicans had used different cerebral areas depending on whether it was a political personality or a neutral personality: this shows there are neuronal circuits *qualitatively* different for motivated and for neutral reasoning – in other words when there are no emotional ties to the subject.

Our reasoning is often influenced by our culture, our personal history, our beliefs, even when we tackle topics that seem universal and unquestionable at first. This is the case for example when we address moral issues such as incest. We will immediately and spontaneously condemn an instance of incest without being able to clearly explain the reasons pushing us to feel disgust. To demonstrate this, Jonathan Haidt,[27] a researcher specialized in moral and ethical matters in New York, did the following experiment. He gathered several social psychologist colleagues of his, and gave them the following situation: Julie and Mark are siblings and travel together during the summer holidays. One night, while they're alone in a hut near a beach, they decide it would be fun to have sex. Julie is on the pill but Mark uses a condom to be sure Julie won't get pregnant. Both take pleasure in it but agree not to sleep together again: they prefer to keep this night a secret that strengthens their ties. Haidt then asks his colleagues what they think of this situation and of the absence of remorse for Julie and Mark. He then observes they are in a state of "moral dumbfounding": all find the situation repulsive and condemnable without managing to justify their moral choice. One could almost speak of motivated morals here.

Interpersonal bubbles and misinformation

This motivated way of perceiving the world can be dangerous if we don't try from time to time to open our mind to contradiction. It's important to be vigilant with the ambiguity reduction we do, above all concerning topics we are subjectively attached to.

In an era of social media and non-stop news, we have access to a plethora of information on every possible subject. It's therefore easy to find material that supports our reasoning and that comforts our beliefs. This goes even further: on social media, we "follow" people who think like us and we look at their posts and news first, which reinforces our own convictions.

Interpersonal "bubbles" are thus formed, particularly on very ideologically sensitive subjects such as politics, religion, veganism, GMOs, protest movements such as the *gilets jaunes* in France, etc. This leads to an even greater polarization of our society and reduces our mental flexibility – in other words, our aptitude to change our mind and our capacity to integrate any new information we're exposed to in the least biased way possible. If I'm pro-*gilets jaunes*, I'm only going to see police brutality and omit there are also violent acts perpetrated by people from my group. On the contrary, if I'm against the *gilets jaunes*, I'm going to focus on the violence of the rioters and demonstrators without seeing that a large part of them are pacifists and that the police is sometimes violent.

When a piece of information confirms our beliefs, we don't really look to know whether it's truthful or not and are then more likely to share it, thus spreading potentially fake news. If, for example, I don't believe in global warming because I think that climate and weather are one and the same, I will be more inclined to retweet former US president Donald Trump's critique of his opponent Amy Klobuchar on 10th February 2019, claiming that she was "talking proudly of fighting global warming while standing in a virtual blizzard of snow, ice and freezing temperatures. Bad timing!" Without even getting into how fake news is transmitted, because we will look at this in more detail in the final part of this book, we can give quite a comical example borrowed from the political sphere, which shows that we are pushed to transmit information that confirms our preconceptions without verifying their source and validity. In 2014, French politician Christine Boutin (former leader of the conservative Christian Democratic Party) retweeted an article published on French satirical website *Le Gorafi* regarding the Family Law Act: "Family Law Act – the government refuses to speak of 'taking a step back' but prefers a 'provisional move forward with deferred potentiality'", attracting considerable mockery. If the origin of fake news sometimes aims to manipulate and disinform the public opinion, its propagation is often carried out by well-meaning people who believe it and think they are doing beneficial work by disseminating it. This tendency to only select information that validates our ideas, our opinions and our beliefs is one of the most widespread cognitive biases: the confirmation bias. We see it at play when it comes to politics and religion, but also on lighter topics such as, for example, our horoscope: we are going to focus on elements that confirm what we believe and above all what we want to believe, and merely glide over those that don't concern us, or that annoy us.

It is useful to be aware of these biases and therefore of the traps our brains set for us. But we cannot forget that these biases also have positive effects and often improve our interpersonal relations. Sometimes, thanks to selection bias, we choose only to remember pleasant times with our loved ones, rather than arguments or difficult moments. Similarly, when a friend calls us and we reply, "I was actually just thinking about you," we activate a confirmation bias and then forget all the times we thought of this person without them calling. Without these biases, we'd have more difficulties to create social ties.

One bias may hide another!

When we take an interest in a subject towards which we have an ideologically entrenched view (immigration, the environment, taxes...), we isolate certain elements which go in the direction of what we believe, and therefore activate both a confirmation bias and a selection bias. Let's take the example of political websites, which are characterized by a considerable selection bias. These sites select only part of the information on a given topic in order to defend their ideological view. It is the case for example of so-called "re-information" platforms such as the French sites Fdesouche, LDC-News, Novopress or TVLibertés, which aim to make far-right theories palatable to the public opinion. Beyond the propagandist appearance of these sites, a thorough study by two French researchers, Yannick Cahuzac and Stéphane François,[28] shows that these sites deliberately convey false information, or truthful but truncated ones, as well as truthful information but with added fallacious content. In short, these sites make the news say whatever they want it to say with the blatant objective to recruit through misinformation. And thus a far-right interpersonal bubble is created online, which we in France call today the *fachosphère*** (from the French slang for "fascist").

Selection bias, coupled with confirmation bias, is a lever also used by influencers on social media. Most of them choose only to show what their followers demand to see: luxury hotels, paradise beaches, makeup and perfect bodies... Recently, more and more influencers, following the Australian Essena O'Neill,[29] denounced what they call the "fake life" shown on Instagram. This "fantasy life" is a source of frustration for their followers because the photos feed the feeling of mediocrity they have regarding their own lives, and for the influencers, this life – which requires them to exclusively

---

*   This does not mean that other media always tell the truth. Only that say fewer fallacies.

post photos of the same standard as the ones before, meaning to never put on a single pound or show any imperfections – creates huge pressure. Everyone ends up losing: the influencers and the followers.

* * *

Even as we become aware of the tricks our brains play on us, it's difficult to be rational all the time, to approach everything like a detective, and to always think objectively. This is partly due to a tension between our beliefs and opinions and, on the other hand, contradictory information. This tension is called cognitive dissonance.

6

COGNITIVE DISSONANCE

*"It's easier to fool people than it is to convince
them that they have been fooled."*
Anonymous (wrongly attributed
to Mark Twain)

To keep motor and cerebral functions in optimum
working order, all living organisms attempt to reach
a state of internal equilibrium called homeostasis.
Human beings don't escape this rule.

You've run a marathon in full summer, and you cross
the finish line dehydrated. In order to regain its homeostasis, your brain is going to send signals to several parts
of your body: to your kidneys to produce less urine,
to your pores to reduce your perspiration, and to your
saliva glands to slow down their activity, because they
are the ones giving you a thirst sensation.

This state of equilibrium is as desirable for cognitive
functions as it is for our body: when information con-
tradicts your preferences, your convictions, your beliefs
or your behaviours, you feel a state of tension which
breaks your homeostasis. This state was theorized by

Leon Festiger,[30] an American psychologist, sixty years ago. He named it cognitive dissonance. In his works, he explains that, naturally, the brain aspires to want to reduce this tension.

Jean de la Fontaine's fable 'The Fox and the Grapes' gives us quite a charming description of cognitive dissonance:

A fox from Gascony, others say Normandy,
Almost starving to death, spied high up a vine
Grapes that were ripe – apparently –
And with their skin the hue of wine...

The fancier would have gladly made his meal of it
But he could not get up to it.
"They are too green," says he, "only fit for a tit."
None better than to moan of it?

The fox is hungry; he would give anything to be able to eat a few grapes. His inability to seize the ripe grapes generates a tension in him, so he enters a state of cognitive dissonance. In order to regain his cognitive homeostasis, he modifies the value attributed to the grapes and changes his opinion to align it with his incapacity to reach them: while they are altogether ripe, he convinces himself that they are too green and that he doesn't want to eat sour grapes. La Fontaine ends his fable with an important question: is this irrationality harmful or beneficial, knowing it has enabled him to resolve his internal tension and avoid frustration?

The same dynamic is at play with smokers: today they cannot ignore that cigarettes are a scourge that stains teeth, contributes to chronic bronchitis, increases the risks of lung cancer, infertility and cardiovascular disease... So to resolve this dissonance, a smoker is going to make up *ad hoc* justifications such as: "I'm too stressed out right now", "smoking is helping me with not getting fat", "I'm too young to have cancer", etc., and will continue to smoke until the dissonance is too great and thus forces them to stop (for example, during a pregnancy or following the smoking-related death of a loved one).

In a book titled *When Prophecy Fails*, published in 1956, Festinger shares his experience of a ceremony in an apocalyptic cult he had successfully infiltrated: Dorothy Martin – mentioned in the book under the pseudonym of Marian Keech – pretended she had received a message from aliens warning her that the end of the world would take place on the 21st of December 1954. Convinced of the veracity of this prophecy, she had successfully manipulated a group of believers, ready to leave everything and follow her in anticipation of their departure aboard a flying saucepan that would come to save them on the day of the apocalypse.

For months before and after the fateful date, Festinger observed the group from the inside. On D-Day, nothing happened. Marian Keech announced then to her disciples that the earth had actually been spared thanks to the forces of "good and light" that the cult had managed to spread around the world.

Festinger recalls that, at that moment, something surprising happened among the disciples: far from abandoning their faith, they launched themselves into relentless proselytism work. With these individuals who invested huge amounts of time, money and emotions into the sect and their belief in a scheduled end of the world, the prophecy's failure created a state of great cognitive dissonance. In order to regain an equilibrium, they preferred to rationalize the failure by convincing themselves that it was thanks to their actions that the catastrophe had been prevented, rather than accepting that they had been fooled. They fabricated and invented a coherent story out of thin air in order to reassure themselves that they had made the right choice when committing to Marian Keech's cult. The world not collapsing strengthened their belief instead of weakening it.

This example is extreme, but it led Festinger to study cognitive dissonance in a lab. In the context of an experiment, he invited human guinea pigs to individually carry out a monotonous and extremely boring task. He placed square cones on a table, then called in the participants one by one, and asked each to pivot the cones by a quarter of a turn, during a whole hour, without giving any further explanations on the aim of this operation.

At the end of the hour, Festinger said each time that the experiment was over, but that he would need a little favour. He pretended that his assistant was absent and asked each participant who had just finished to go tell the next one (who was in fact his assistant) that the experiment was a very enjoyable experience. Festinger offered to pay for this service: he offered one dollar to half of the participants, and twenty dollars to the other half.

All participants played the game. After paying them, Festinger asked them individually if they really, without lying, found the experiment as enjoyable as that. At a first glance, one could think that those who received the most money pretended to find the experiment enjoyable, yet it was the participants who were paid one dollar who pretended to enjoy the experiment the most. With the group that paid twenty dollars, the financial reward was enough of an added piece of information to compensate the time lost during the experiment and reduce the dissonance. The best-paid guinea pigs tell themselves: "the task was boring but at least I was paid well – it's useless to justify it with any interest on my part". On the contrary, those who were paid one dollar have had to modify their perception of the experiment to rationalize losing one hour of their life. They started to sincerely believe it was interesting to participate in a scientific experiment, as repetitive as it may have been.

Festinger defined three steps in the reduction of the cognitive dissonance process: first, identify the event which makes us enter into dissonance, then, adjust our behaviour or belief to regain consonance and finally, if necessary, add new information which can reduce the effects of the dissonance (here, the salary for the service rendered).

The reduction of this cognitive dissonance is a phenomenon which is frequently carried out in our daily life and which shows us once more how capable we are of distorting reality to reconcile our ideas and our behaviour: the same cognitive dissonance reduction mechanism at play with smokers is triggered when you keep eating meat even though you are aware of the terrible conditions in which the animals are reared and killed, as well as the carbon footprint producing this meat has, or when you keep buying clothes from mass retail brand X or Y knowing that they are manufactured in extremely precarious working conditions.

Manipulating others through cognitive dissonance

Once we are aware of this mechanism, it is possible to use it intentionally towards others. In his autobiography, Benjamin Franklin recalls how he handled his relationship with a pugnacious political rival while he was in power. Knowing that this man was a fervent rare book collector, Franklin one day sent him a letter in which he asked to borrow a selection from his library, particularly "a very scarce and curious book". One can imagine Franklin's rival entering dissonance following this unexpected request, torn between the negative opinion he had of Franklin and the fact the latter had asked to borrow a book from him. Three ways to reduce the dissonance presented themselves to him then:

- Trivialize the event. Franklin's rival could tell himself the request was insignificant. But it was impossible to consider a letter from Benjamin Franklin insignificant at the time.
- Add new information that could help him reconcile his contradictory beliefs. For example, asking Franklin to send him books in return – but everyone knew that Franklin wasn't a bibliophile.
- Modify his behaviour or his belief towards the dissonant factor: as he couldn't refuse to send the book to Franklin without being ridiculed by everyone,

the only way for him to reduce the dissonance was to modify his initial belief: he could only consider Benjamin Franklin in a more positive light and get the requested books to him at once.

A week later, Franklin returned the books to him after having slipped a thank-you note in one of them. When the assembly reconvened, Franklin's rival addressed him directly for the first time and thanked him for his kind note. Benjamin Franklin relates that after this, he "ever after manifested a readiness to serve me on all occasions". They even became great friends, and their friendship continued until his death.[31]

One might thus believe that someone who's already done you a favour will be more inclined to do more. In other words, we don't only serve those we appreciate, we also appreciate those we serve. We adapt our actions and our judgements to someone else according to the way we interact.

What we call the "Ben Franklin effect" is used today in the business domain. If you buy an iPhone, the most expensive phone on the market, you cannot consider it to be an average phone. Even if there are sturdier, faster or more aesthetically pleasing ones, iPhone users can't conceive the existence of better ones since they have bought the most expensive of them all. It would be admitting they've been "fleeced". The whole luxury and contemporary art industries operate according to this monetary commitment principle. The more we pay the more we feel we have bought something truly luxurious. If we buy a bag for £200, we don't feel like we have bought as luxurious a bag as we would if we treated ourselves to a £35,000 Birkin, while the difference in quality might not be huge.

Using dissonance mechanisms towards positive ends

Cognitive dissonance also has its positive side: it can be used as a tool to handle the stress situations mentioned earlier. If you're stressed by the idea of speaking in public in a meeting, and you anticipate the stress this situation is going to cause, you will have a tendency to avoid it rather than face it. By staying at home, you will keep your balance, your homeostasis. Choosing to avoid the stressful situation validates your anxiety: if you're not going to the meeting, you're reinforcing the idea of a real danger. It's in order to break this vicious circle that we need to intentionally enter dissonance.

In therapy, we use a technique which consists in a patient's gradual exposure of the stress-inducing object or situation to the patient. The aim is to trigger a dissonance between a belief that whispers "it's dangerous" and a behaviour that incites them to go for it anyway. To resolve this dissonance, the patient will have to modify their fateful belief. They will then think: "If I'm going for it, it means that it's not that dangerous", and they will gradually regain balance without having to use avoidance. This is gradual exposure therapy.

Cognitive dissonance can also help us to congratulate ourselves on a choice we have made. After hesitating between two objects, we tend to overestimate the object we ended up picking, and to underestimate the object we left behind. You want to buy a car, and at your dealership you don't have one but two favourites: two cars completely different from one another. Both are in your budget and you have to pick one. Imagine a friend who's also wondering which car to buy asks you which one of the two cars you like the most. You will reply that you prefer the one you have bought, while initially, you didn't really have a preference at all between the two. You will lie to yourself to avoid the discomfort we feel when we have to give up something we like, as our fox with the grapes. We constantly change the value of the information that reaches us and with which we interact through mechanisms of cognitive dissonance reduction.

When we are blinded by an excess of coherence

We're always looking for more coherence. But is it possible, or even desirable, to have a completely coherent image of oneself? The Myers-Briggs personality test example can help us to find an answer. This test was developed by Katherine Cook Briggs and her daughter, who intuitively came up with the theory that there were several wide universal personality types in the world.

In 1944, they published their first version of the MBTI in a book titled *The Briggs Myers Type Indicator Handbook*. Then, in 1956, they published the *Myers-Briggs Type Indicator*, which gave the test its official name. This test consists of a series of about ninety closed-ended questions for which the participants have the choice between two answers. After having answered, a participant receives their profile type among sixteen possible combinations.

Today, the MBTI leads the corporate personality test market. It is used by close to two million people in the world each year and brings close to twenty-million US dollars per year to the company selling it, according to a joint investigation by *Le Figaro* and the *Washington Post*.[33]

Often used by human resources as a career orientation tool and a performance predictor in recruitment, we find in this test tables indicating which types of jobs best match such and such a personality. We know, however, that this test isn't reliable, because if we give it to the same person several times, we obtain results that are different, even contradictory.[34] How did this test become essential in the corporate world when it was never validated by a competent authority, isn't really reliable and has no theoretical basis?

To answer this question, we need to introduce a psychological phenomenon called the Barnum or Forer Effect. In 1949, Bertram Forer, a psychology professor, subjects the thirty-nine students in his introductory course on psychology to a test, telling them this tool will give them a brief overview of their personality. One week later, he hands each test student their respective results, asking them to check if the result is true to their personality. The students don't know that they have all received the same result, made up out of bogus sentences borrowed from a horoscope. This is what it says:

1. You have a great need for other people to like and admire you.
2. You have a tendency to be critical of yourself.
3. You have a great deal of unused capacity which you have not turned to your advantage.
4. While you have some personality weaknesses, you are generally able to compensate for them.
5. Your sexual adjustment has presented problems for you.
6. Disciplined and self-controlled outside, you tend to be worrisome and insecure inside.

7. At times you have serious doubts as to whether you have made the right decision or done the right thing.

8. You prefer a certain amount of change and variety and become dissatisfied when hemmed in by restrictions and limitations.

9. You pride yourself as an independent thinker and do not accept others' statements without satisfactory proof.

10. You have found it unwise to be too frank in revealing yourself to others.

11. At times you are extroverted, affable, sociable, while at other times you are introverted, wary, reserved.

12. Some of your aspirations tend to be pretty unrealistic.

13. Security is one of your major goals in life.

Forer then asked his students to raise their hand if they were satisfied with the test results: almost all hands were raised. Impassive, Forer began to read the first answer, then the second before the entire class burst out laughing, understanding the deception.

The Barnum Effect is therefore a bias that leads us to believe a statement that says something about our personality, and this due to three factors: because we think that the statement had been written especially for us (personalization bias); because the person addressing us is an authority figure (authority bias); finally because the statement is vague and general enough while being sufficiently positive to make us want to believe in it (selection bias). We understand why, in the case of personality tests that can cost a fortune to companies or individuals, we can talk about scams: these tests subsume three biases into one.

* * *

Beyond personality tests, motivated reasoning and the reduction of cognitive dissonance are at work in practically everything we undertake. The important thing is to be aware of it, and to ask ourselves why our action or cognition influences at any given moment the value we attribute to the things around us, our social relationships and our opinions.

The question asked at this point is the following: until now, we have considered ourselves *actors* of our life while accepting that we sometimes make mistakes. But are we always active beings? Might we not be confronted with a loss of control?

7

WHAT I CAN CONTROL AND WHAT ELUDES ME

"What depends on us must be perfect; as for the other things, take them as they come."
Epictetus

We can say that there are basically two types of people: those who think they only owe to themselves whatever comes their way, and those who, like Diderot's Jacques the Fatalist, think that "everything has been written up above" ("up above" can be related to any type of transcendence here). It's the starting point for American psychologist Julian Rotter[36] in the elaboration his social learning theory.

Julian Rotter calls locus of control these two ways of approaching life: those who believe that events only depend on them have an internal locus of control[37] (ILC) and those who think that events are the deed of external factors have an external locus of control (ELC). If at work you obtain a promotion and that you have an ILC, you will tell yourself it's thanks to your own efforts. If on the contrary you have an ELC, you'll attribute this promotion to luck or an absence of competitors. Let's specify now that the locus of control isn't a binary variable: we never have a 100 per cent internal or 100 per cent external locus, but we lean towards one rather than the other, and this can evolve throughout our life and depending on our experiences.

In 1955, Jerry Phares, one of Rotter's students, put an experimental protocol in place to show that our locus of control has an important impact on our performances and our self-esteem. He entrusted two groups of people with carrying out the same, very easy task,[38] which consisted in determining among several angles which ones were equal. He told the first group that this task was very difficult and depended on a great deal of luck (ELC), and to the second group that it was a matter or competence (ILC). Phares then asked the participants from the two groups to evaluate the percentage of chances they had of succeeding or failing. The results obtained showed that their way of self-evaluating a priori was conditioned by what they believed to be or not to be up to them. Those who thought it was about luck were less sure of succeeding in the task than those who thought their success only depended on their own competence.

On a daily basis, we don't always have someone behind us like Phares to tell us that one thing depends on us and another one is a matter of luck. It's up to us to evaluate how much control we have on our actions. And when we evaluate poorly, this can have significant consequences. For example, the idea that women are worse at mathematics than men is widely spread, while a biological difference between genders which would justify a difference in mathematics capacity levels has never been observed. It is what we call a negative stereotype bias: we unconsciously attribute negative characteristics to a population group without any foundations.

Steven J. Spencer,[39] a psychology professor at Ohio State University, wanted to see if it was possible to remove this bias through the following experiment: first of all, he gathered a group of men and women with a similar proficiency in mathematics, and he subjected them to a standard exam. The result of this test indicated that the men had a greater success with the exam than the women. Afterwards, he repeated the same experiment with two mixed groups, with one difference, however: the first group was told that the test had been done before and that men had greater success than women. In contrast, he told the second group that the results of the previous tests indicated that men and women had similar performances. Let's remember that the two groups took the exact same test.

Logically, if the performances were due to innate differences between men and women, the results of the second experiment should be similar to the results of the first one. Yet, while the results of the first group indeed showed a big level difference between men and women, the results of the second group indicated that men and women had quasi-identical success rates. One sentence was enough to remove the difference in performance. The results of the second group show that it is therefore possible to "recalibrate" the women's locus of control towards a more internal axle by suppressing the societal stereotype bias (which for women isn't limited to mathematics).

Locus of control and feelings of responsibility

Our locus of control conditions our actions and therefore the results of our actions. We don't feel responsible of them in the same way whether we lean towards an ILC or an ELC. Numerous research articles[40] have indeed demonstrated that an ILC leads the subject to more self-accountability and contributes to greater self-esteem than an ELC. If you lean towards an ILC, you will attribute your successes to your own skills and you will develop a good self-esteem. The more you lean towards an ELC, the more you will attribute your victories to external factors and therefore get less personal satisfaction from them. In the case of a failure, if you lean towards an ILC, you will tell yourself that you will do everything to succeed the next time, while if you're leaning towards an ELC, you will be more fatalistic and you will have a tendency to attribute your failures to factors that you can't and will never control.

However, leaning too much towards an ILC can also have negative consequences. In the context of events that do not depend on you (or at least not uniquely on you), such as being made redundant on economic grounds, you will internalize the failure and react as if what has happened to you were your fault, which can lead you to develop symptoms of anxiety or even depression. On the other hand, in this type of situation, someone who rather leans towards an ELC will put things into perspective more and will overcome the shock caused by the redundancy with more ease.

Now let's examine the impact our locus of control has on our health.[41] If someone suffering from cancer feels responsible of their illness or if, on the contrary, they believe that same illness is solely a matter of bad luck, they will not react the same way to the diagnosis and to the treatment. The way we consider our health shapes the way we face illness.[42] A patient who leans towards an ILC will probably put more energy in their healing process and will be more conscientious in their medication intake than a sick person who believes that, whatever they do, 'everything has been written up above'. The latter will adopt a more passive attitude towards the disease, an attitude of learned helplessness, which could be summarized thus: if nothing depends on me, what's the good in fighting and trying to get through this?

Learned helplessness

The American researcher Martin Seligman led experiments in an attempt to explain how syndromes of learned helplessness can develop following trauma. He conducted the following experiment on two dogs. Each dog was in a cage with a floor that could transmit electric shocks. In each cage there was a small lever which the dogs could press.[43]

Seligman sent a first shock to the two dogs. Before each shock, a little lamp was switched on. At the moment of the shock, both dogs first attempted to flee, then tried their luck with the lever. In cage 1, the lever worked and the shocks ceased immediately. In cage 2, the lever was fake, and shocks continued even if the dog pressed on it. After several shocks, Seligman observed that the dog in cage 1 pre-empted the electric shock and jumped on the lever when the light was switched on, while the other dog progressively abandoned and lied down on the ground for each shock: it had understood that it couldn't change anything about this.

Next, Seligman put each dog in a double cage: a small wall split the cage into one half with an electrified floor and the other half without an electrified one. Seligman began the experiment again. Dog 1 tried to cross the small wall to see if this saved it from getting electrocuted and succeeded in avoiding the shock. Dog 2, even if it now had the opportunity to act, didn't attempt anything: it automatically lied down as soon as it saw the incoming shock. Learned helplessness is therefore an individual's incapacity to escape an adverse situation even if they have the opportunity to do so.

Donald Hiroto, a colleague of Seligman's, wanted to see how learned helplessness was expressed when it came to humans, and recreated Seligman's experiment replacing dogs with humans, and the electric shock with a detonation.[44] As previously, the experiment involved two participants, one with the power to stop the noise and the other without, and took place in two stages. The results were the same as with dogs: the participant who understood that they could stop the noise by pressing a button four times in a row succeeded in stopping it easily, even in the second part of the experiment. On the other hand, the second participant who had no control over the noise to begin with showed similar reactions to the ones Seligman had described with the dogs, and didn't even try to press the button in the second part of the experiment, when they could have stopped the sound.

With human beings, learned helplessness is often linked to depression. When we are in the midst of a depressive episode we feel that we no longer have any control over our own life. Learned helplessness also manifests itself in other terrible situations, such as domestic abuse.[45] The traumatic experiences, i.e. the quasi-daily violence, makes the victim develop an external locus of control to an extreme level: they end up convinced that it's impossible for them to leave their spouse, or even to do anything to stop the violence and get out of the toxic relationship. In fact, repeated assaults make the victim enter a state of dissonance because they have done nothing wrong, which can push them to develop a feeling of guilt in addition to learned helplessness, in order to justify those assaults and regain coherence: "If I'm beaten up, it's my fault, so really, it's normal." They will also attempt to rationalize the assault by making it acceptable in order to subconsciously justify their lack of reaction: "He/she doesn't do it on purpose; he/she loves me underneath it all."

Aside from the physical and mental pain caused by daily abuse, learned helplessness will also have serious consequences on the professional performance or the social relationships of the victim, because it drastically reduces their self-esteem. The domestic abuse issue is obviously very complex and multifactorial, and doesn't only come down to learned helplessness or cognitive dissonance reduction. But they are important elements to take into consideration in order to grasp how difficult it is for a woman (or a man) who is the victim of domestic abuse to take the step of leaving the family home and thus the abusive person, or even call a helpline for victims.

Learned helplessness also plays a role in societal issues such as global warming.[46] Indeed, researchers put out the hypothesis of humankind's learned helplessness and consequently quasi-generalized inaction in the face of global warming which would lead to the end of the world as we know it— and maybe even to the end of our species full stop. The fact that, on a personal scale, individuals consider that their actions cannot have a real impact on the climate translates into a general demoralization which hinders any environmental action. We have more and more access to studies[47] proving that large-scale action on climate is only possible if each individual does their bit: it's up to us alone to begin to fight against generalized learned helplessness, overcome our apathy and finally act. This does not of course remove the need to modify our systems of governance on a global scale and move towards a system that is more respectful of nature.

Other cognitive biases will also work towards preventing us to act, without us even being aware of it. For example, in the case of global warming, the present bias makes it difficult for us to project ourselves into the future. In fact, we are a lot more sensitive to events with immediate and visible consequences than to a more distant future. If we had even only one "chance" out of ten to develop cancer thirty minutes after smoking a cigarette, there probably wouldn't be any smokers in the world. In the global warming context, this translates into a difficulty to picture ourselves a few decades from now. When you are told that, in a hundred years, the ice floes will have melted, you don't feel that it concerns you because it is difficult to imagine it. So why worry about it now?

Another problem: many companies play down the impact our polluting activities have on global warming and justify it by saying that in any case, by then, a technological solution will have been found. Numerous companies consider that it is not up to them to solve the climate issue, but up to governments, politicians or other companies. And so they reject all responsibility for the visible degradation of our environment. It is what is called a diffusion of responsibility bias. We will come back to it in more detail in Chapter 9.

Learned helplessness bias plays a role in a multitude of less serious situations, ranging from the lack of courage to leave a job to not being bothered to go vote… We have seen that too external a locus of control leads to learned helplessness and to situations that can be very serious, but how about a locus of control that is too internal?

The illusion of control

If we trust examples of cases where an external locus of control inhibits our action and hinders our willpower and our free will, it seems that, in comparison, an internal locus of control is a good thing. Nonetheless, leaning towards a locus of control that is too internal leads to an illusion of control and self-control which can have harmful consequences on our mental health and that of those around us.

Someone who thinks they can control everything will tend to blame themselves for a mistake they haven't directly made, and will be exactly as intransigent towards others, considering them as capable of the same absolute control. It's not for nothing that we often use the expression 'control freak' to talk about it.

People with a very internal locus of control are very often perfectionists in the negative sense of the term, which makes them fall more easily into the binary thinking trap mentioned in Chapter 3. A control freak will think that everything that is not absolutely[48] perfect is completely rubbish and only worth throwing away. They lose their ability to perceive nuance, which is necessary to the elaboration of critical reasoning, and they reach a form of mental rigidity. This is often (mis)taken for arrogance, which makes people with an excessive ILC struggle to have healthy and fulfilling social relationships.

Another pitfall: someone with an excessively internal locus of control can develop a "saviour syndrome". They will want to solve other people's problems to the point of being intrusive, refusing to let them make mistakes, learn and find their own path: we can't "save" someone against their will!

An excessively internal locus of control can be a real hindrance to being a part of society: when you think you can control everything by yourself and you are faced with an emergency, you will hesitate to ask for help since, for you, everything is under your control. Similarly, if during your studies or at work you are led to work in a team, you won't manage to delegate, not even to let your teammates do their part of the job.[49]

Believing we don't have a grip on our environment and sinking into inaction or even apathy is a trap. Believing we're omnipotent and that everything depends on our willpower is also a trap. Therefore, there isn't an inherently good or bad locus of control. The importance is to not lean too much towards one or the other. The only way to find this balance is to analyse situations as best we can in order to determine how much or how little things depend on us. But this knowledge of the world and of situations is far from easy, especially because we are often the victims of an illusion of knowledge.

8

THE ILLUSION OF KNOWLEDGE

"A little learning is a dangerous thing; drink deep, or taste not the Pierian spring: their shallow draughts intoxicate the brain, and drinking largely sobers us again."
Alexander Pope

On the 6th of January 1995, MacArthur Wheeler robs two Pittsburgh banks one after the other, his face completely bare. He's arrested in April, and when the police tells him he's been recognized thanks to CCTV, he's flabbergasted and says, "But I wore the juice!" Wheeler then explains that someone had shown him how to make invisible ink with lemon juice and he had thought that if he coated his face with lemon juice his image wouldn't appear on security cameras. He has convinced himself of the viability of his idea by taking a Polaroid selfie while his face was smeared in lemon juice: he didn't appear on the photo! To explain this phenomenon, the police assumed that his camera was faulty or that he simply hadn't aimed properly. Whatever the reason, Wheeler paid the price of his illusion of knowledge and was sent to prison a few days later.

This news story features in the 1996 *International Almanac*, where David Dunning, a Cornell University psychology professor, found out about it. Reading about Wheeler, Dunning imagined this story was quasi-universal: the less we know about a subject matter, the less we are capable of measuring the extent to which we do not master the subject at hand. You may have been faced with this type of situation: you have friends over and you decide to recreate a recipe you saw on a cooking show the night before. After all, it didn't seem that complicated on the screen! You're self-assured, you think your dish will be delicious, you even feel a certain pride serving it to your dinner guests. But it's a cold shower: nobody finishes their plate. You have been the victim of a self-confidence peak when it comes to your culinary competence! A show broadcast on French channel 6Ter called *Norbert, commis d'office* (*Norbert, Appointed by the Court*) is actually dedicated to "taste criminals" who don't master any basic cooking rules but fancy themselves true chefs. The reality TV show's participants state to the camera that they are convinced of their culinary talents and aren't surprised they have been selected by chef Norbert Tarayre to take part in a collective culinary project. When Norbert reveals that a loved one has denounced them for a culinary infringement, they are flabbergasted and don't take the accusation well. The chef then offers to teach them to cook a gastronomical version of the dish they thought they mastered. Discouraged, most of them struggle to get on with it. But under Norbert's leadership they gradually regain their confidence and accomplish a dish worthy of a great restaurant. The show's participants have therefore swayed between an unjustified self-confidence peak, followed by a state of despair emerging from the realization of the breadth of their ignorance, to finally gradually rise up in knowledge and self-confidence.

This can be applied to all parts of life: when we begin to play a musical instrument, we often think that it's not that complicated. On the piano, we can easily play 'Au clair de la lune' after a few minutes only. But if we persevere in learning this instrument, we will soon understand that progress will not be that quick: we will need months, even years to master Beethoven's sonatas. We will go through a phase of confidence loss and demotivation. We will even sometimes think that we'll never succeed: we will then have to go past this plateau through hard work. Similarly, when we learn a new language, Spanish for example, we can get by talking about basic things very quickly. On the other hand, if we open *Don Quixote* in Cervantes's original tongue, there's a strong chance we'll be dizzy at the thought of how much there is still to learn! All learning often starts with an unjustified confidence peak regarding our knowledge on a topic.

Dunning and his student Justin Kruger wanted to give a scientific foundation to this cognitive journey by carrying out several experiments which have enabled them to coin the effect which today bears their name and has the following shape:[51]

Dunning–Kruger Effect

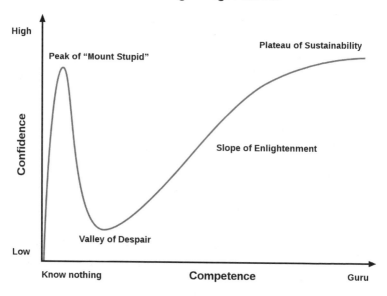

To draw up this curve, the two scientists carried out a first experiment to confirm we indeed have a confidence peak regarding our knowledge when it fails us the most. Dunning and Kruger gathered a panel of students, to whom they asked several grammar and logical reasoning questions. Before communicating the results to them, they asked the participants to evaluate their success rate. The experiment showed that the worst students in the class had the greatest tendency to overestimate their results and their ability. They stopped at their first peak of self-confidence. In the context of a second experiment,[52] Dunning and Kruger wanted to see if it was possible to decrease this self-confidence peak. So they gathered the students who were initially the most self-confident, and shared the answers of those grammar and logical reasoning questions with them. They came to this conclusion: "Paradoxically, by improving the intellectual capacities of the students and by instilling new knowledge to them, we have helped them to realize that they didn't know enough to begin with, and therefore, to become aware of the limit of their knowledge."

They then asked the students what their impressions were following this experiment: by coming down from their confidence peak, and by assessing what remained to be learnt, the students first went through a discouragement phase before understanding they could resume their ascent towards knowledge. This is how the curve on the previous page was made.

This self-confidence peak is explained by our illusion of explanatory depth: we often think we understand the world better than we really do. Inspired by Dunning and Kruger's works, the British researcher Rebecca Lawson[53] wanted to prove that we do not only make mistakes on the depth of our knowledge but also on its relevance. She set up the following experiment to show that we do not fully understand how daily objects work: Rebecca Lawson gathered a sample group of adults who had all cycled before, and asked them to draw a working bicycle from memory. We are reproducing here various examples of drawings carried out by the participants:

(A)   (B)

(C)   (D)

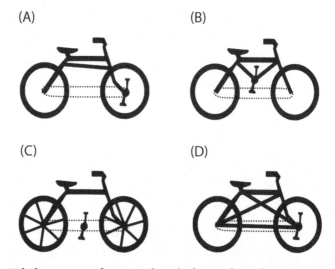

While many of us tend to believe that there is nothing simpler than to draw a bicycle, none of the bikes above are able to function. In total, almost none of the participants managed to draw a working bicycle, and 40 per cent even failed to identify which bike could potentially function among several representations. To visualize something – meaning to see it in its absence and clearly explain how it works – is a much more difficult task than to reproduce what's under our very own eyes.

We constantly overestimate our ability to understand how the world works. What's important is to be aware of it and to not stop at our self-confidence peak each time we discover a new discipline or we're confronted to new ideas. On the contrary, let's agree to dive into the depths of knowledge: we will do it all more willingly, since, as the curve demonstrates, after the discouragement in the face of the scale of what remains to be learned comes the move back up towards more solid knowledge.

Knowledge illusion: societal and political consequences

In their study 'Boss Competence and Worker Well-Being', three British researchers stated a problem we find in many companies today: it is not always the most competent people who get promoted, and underqualified people are often found in important positions.[54] For the people stopping at their self-confidence peak, the Dunning-Kruger effect bestows a real feeling of superpower. Mediocre people will thus dare to aspire to positions for which they are not qualified, and their self-confidence will enable them to get them.

In this book *Les Adultes surdoués* (*High-Potential Adults*), French child psychiatrist Gabriel Wahl[55] explains that, contrary to the above, overly competent people and especially so-called high potential people underestimate their abilities and constantly fear not being "up to it". This syndrome, often called "impostor's syndrome", is the negative counterpart of the Dunning-Kruger effect. It leads people who suffer from it to accept positions beneath the ones they deserve. And combining these two effects leads to an absurd reality: under-skilled people manage over-skilled ones. As early as the 1970s, Laurence J. Peter and Raymond Hull, two Canadian professors specialized in corporate hierarchy issues, fine-tuned this theory in a book called *The Peter Principle*[56] They showed that, in general, within a company, any employee will be promoted up to their incompetence threshold. As they cannot be demoted, they will continue their career with responsibilities they're unable to fulfil. Overconfidence in the less able, impostor's syndrome in the more competent, all of this combined with Peter's Principle, are pitfalls for company directors who wish to avoid their companies being led by senior managers unable to fulfil their responsibilities while more competent people are relegated to subordinate positions.

Anne Boring, Head of the Women in Business Chair at Sciences Po Paris, recently explained that impostor's syndrome is also a real gender issue: "Young women suffer more from impostor's syndrome than men. They don't always feel they can legitimately defend their projects.[57]" The consequence is that women dare less to aspire to senior positions than men, fearing they cannot fulfil those responsibilities, thus creating a troublesome precedent: the less they fill these executive or better-paid positions, the less women from the next generations will dare to do it in turn, also internalizing this learned helplessness. In 1978, American clinical psychologist Pauline Rose Clance was the first to theorize this in a book, *Impostor Phenomenon*, which she wrote after observing a complete inability to internalize success[58] among several women around her. While they were all pursuing brilliant careers, they wouldn't stop putting themselves down. Today still, 44.8 per cent of the female workforce in France is concentrated in lower-income sectors such as public administration, health, education or social work.[59] However in 2017, in France, 31.3 per cent of women aged 25 to 34 had obtained a diploma above BA level, against only 26.4 per cent of men.[60] So while on average they have more diplomas than men, there is still among most women an unconscious acceptance that they have fewer chances of succeeding professionally. Even if today, Western societies have generally evolved towards greater gender parity, even if many women succeed, a woman will less often dare to fight for a position than a man would, fearing she won't be good enough. For the same reasons, she will not ask, with equal skills, for a promotion or a pay rise as often as a man would. Impostor's syndrome is therefore riddled with a form of learned helplessness among women.

When false ideas ring true

We are all subject to self-confidence peaks. It's what makes us sometimes inclined to, (when we don't come down from this peak) take simplistic and false ideas as absolute truths. In the health sector, trends based on primary, spontaneous and seemingly coherent beliefs emerge. Too simplistic an approach for complex subjects, such as the beneficial or harmful effects of vaccination, for example, can lead us to a peak of self-confidence when it comes to our general understanding of this matter. A fringe of the anti-child-vaccination movement is based on the idea that vaccination may be a cause of autism among children. This idea was developed following a fallacious article on the subject matter. In France, in 2016, a study showed that only 59 per cent of French people "still trusted" vaccines.[61] Yale University Researcher Matthew Motta[62] took an interest in this sceptical discourse, held most often on social media by non-experts. Motta realized that this discourse was particularly assertive after the free global broadcast of the film *Vaxxed* between the 1st and the 8th of November 2018. This anti-vaccination documentary seeks to demonstrate the dangerousness of the Measles-Mumps-Rubella vaccine (MMR), the only compulsory vaccine for newborns in France, by showing a series of cases of children who developed autistic spectrum issues after a vaccination. The mistake is a double one: firstly, in *Vaxxed*, no evidence of a causal link between the vaccination and autism is successfully brought forward. Secondly, the film doesn't take into consideration all the other cases (a lot greater in number) in which children were vaccinated without developing disorders of the kind. To sum this up, the outspoken critics of child-vaccination choose anecdotal evidence ("I know someone who…") rather than scientific proof ("X% of people have developed a disorder…"). Motta thus noticed that most anti-vaccine militants, after seeing *Vaxxed*, were convinced they mastered the topic better than physicians themselves. This proves their overconfidence bias. Following the increasingly popular adherence to this movement, which goes beyond the MMR vaccine (see for example the opposition to COVID-19 vaccines), a decrease in the vaccination rate was observed,[63] leading to a resurgence of diseases such as measles, tuberculosis or scabies among children, which we believed had completely disappeared.* Here lies the danger of believing in theories founded on too simplistic an explanation.

---

* Here we're specifically speaking about the false belief in the existence of a link between autism and vaccination. We do not want to say that a vaccine has never had any negative effect on someone's health, as it wouldn't be true. We only want to show that nuance is of the essence and that we must not fall into the trap of false equivalences.

Donald Trump was perching on a self-confidence peak of his own when, for example, he introduced a theory justifying his abstinence from any form of physical exercise. Two *Washington Post*[64] journalists reported that Donald Trump had observed that many of his sportiest peers at university had had health issues later on. Out of this observation he deducted a false correlation: the human body may be made of a limited quantity of energy – like an electronic device – and sport may contribute to using up this battery. He thus concluded that, to remain in good health, one shouldn't do any exercise!

Nowadays, people want us to believe that everyone can understand everything very quickly. On YouTube, videos promise to teach us everything on any complex subject (politics, science, ecology...) in a few minutes flat. "Expert" blogs on all imaginable subjects suddenly appear each day and sometimes meet a lot of success. By giving a voice but also an audience to anyone, the internet has been the best support for the emergence of false experts and real charlatans. In the words of Umberto Eco, "Social media gives legions of idiots the right to speak when they once only spoke at a bar after a glass of wine, without harming the community [...] but now they have the same right to speak as a Nobel Prize winner."

This profusion of information sometimes leads us to make choices which give us the illusion of knowing the whys and wherefores of an issue. In England, on the 24th of June 2016, the day after the Brexit referendum, the no. 1 search term on UK Google was "What is Brexit?", and in second place was "What's the European Union?".[65] This shows that, on the 23rd, many British people had voted without really knowing what they were voting for or against. Once faced with the ballot results, they wanted information on it. This does not stop here: today, many "Brexiteers" regret their vote and are asking for a new one. In July 2018, a petition for a new referendum launched by the daily newspaper *The Independent*, on then Secretary of State for Education Justine Greening's initiative, gathered 200,000 signatures[61] in two days only. In November of the same year, the company Survation organized a poll to know if British people were for or against the UK's exit from the EU: 54 per cent voted against Brexit, while 54 per cent of British people had voted "Leave" in June 2016. Now better informed on Brexit, some Brits seemed to want to go back on their original choice.

Simplification traps and pseudo-profound bullshit

Many complex issues can be wilfully reduced to a false simplicity: this process is part of a trend called "pseudo-profound bullshit".[66] American philosopher Harry Frankfurt determines that bullshit is "speech intended to persuade without regard for truth".[67] In contrast to lies, which are a deliberate manipulation of the truth, bullshit leans on false or reductive spontaneous beliefs by wrapping it all up in attractive speech. The idea is to profit from it, often financially.

Topics linked to health are the boon of the pseudo-profound bullshit industry. The detox industry for example coats their "slimming programmes" and other "body detox cures" in beautiful promises and reaps great financial reward from it. Rather than making the decision to eat and live more healthily, we sometimes prefer to let ourselves be seduced by the pretty packaging of detox juice cures advocated by our favourite celebrities, or trust slimming dietary supplements which won't miraculously help us to slim down if we continue not to feed ourselves properly between two detoxes and don't do any physical exercise.

In the same way, some personal development trends make beautiful promises, so well phrased that they seem full of depth, while they're often empty of meaning. To show that it's sometimes difficult to detect

120

their vacuousness, University of Regina psychology professor Gordon Pennycook[68] carried out the following experiment: he artificially and randomly created ten pseudo-profound phrases using two websites. The first website makes a list of the most commonly used words in the tweets of Deepak Chopra, a famous self-help book author and a fervent defender of alternative medicines. The second site bears the unequivocal name "New Age Bullshit Generator": it's a site that uses a list of words that are recurrent in esoteric quotes and then randomly combines them. Here is one of the phrases generated through this process: "Hidden meanings transform the abstract into the beautiful." Or also: "We are the brothers and sisters of the infinite." Pennycook then gathers a panel of Chopra readers and of adepts of self-help methods. He reads ten artificially generated phrases in the middle of authentic Chopra quotations and asks them to decide which ones have been artificially created. The participants can't do it. Indeed, every sentence is vague enough to be given some sense and be interpreted as one wishes – a bit like with a horoscope.

But even if these sentences are devoid of meaning, what's the harm in believing them? Don't we have the right to bring a little bit of poetry to our lives, and a little optimism? The problem isn't with these sentences in themselves, but in the cases where this pseudo-profound rhetoric presupposes "therapeutic" methods presented to be as effective as traditional medicine, when gurus play doctor. I attended the conference of a fellow psychologist who was telling the story of one of his patients, an adept of Chopra and of his Ayurveda-inspired[69] therapeutic methods. For seven years, this mother really believed in their efficacy and never took her children to a "real" doctor. One day, one of her sons became ill: as usual, she gave him Ayurvedic treatments, without success. The boy had a *Staphylococcus aureus* which can only be cured with antibiotics. When he arrived at the hospital, it was too late, and one of his legs had to be amputated. The lesson to be drawn from this is the following: we can let ourselves be seduced by alternative methods adorned with pseudo-profound waffle – as long as we are not truly ill, because they can actually work as a placebo.* But in the case of a serious illness, whether it's physical or mental, too great a trust in these shamanistic medicines can put us in danger.

Today these methods are popular, and just as it is with detox trends, there's a real self-help industry out there. Some books that sell hundreds of thousands of copies are built on attractive promises: to succeed in life, become rich, find true love, etc., while some

---

* In the collective unconscious, the placebo is often considered to be something ineffective. It's not true. A placebo can be effective in many cases – which doesn't however make it a remedy.

lucky readers do succeed – and credit their success to these books while they above all owe it to themselves – they are not the majority. If the millions of readers of Rhonda Byrne's bestselling book *The Secret* had successfully accomplished what this self-help method promises, we would know. The book begins like this: "By becoming aware of the Secret, you will discover how you can have, be or do everything you want." And to reach this goal, Rhonda Byrne offers an infallible "scientific" method: the law of attraction. Here are its pillars: if you think of something you dream of (love, money, success...), you will be able to attract it thanks to the electricity contained in your brain, which will create a magnetic field and attract the universe's positive waves. Do you want to become rich without working? No problem: sleep with a banknote on your forehead!

This theory has no scientific or medical foundations. However, some treatment centres, such as JMC Psychotherapy in the US, ground all their therapeutic methods on the law of attraction. Thanks to those, they guarantee they will treat severe trauma linked to, for example, sexual abuse, and pretend they heal certain mental illnesses such as depression or eating disorders. There are also "certificate training courses" for becoming a "law of attraction master coach". A patient recently told me she had participated in a seminar during which the speaker had wanted to prove the legitimacy of the law of attraction by asking the participants to stick a one-euro coin to their forehead. When they had successfully completed this exercise, the speaker then explained that the brain acted on the coin like a magnet, thanks to the infamous magnetic field. *Bullshit*: if the coin sticks to the forehead it's because it sticks to the skin (the effect would have been the same if he had tried to stick it on his leg, his belly, etc.). Nothing to do with a cerebral magnetic field – otherwise our forehead, like a magnet, would be attracted by all fridges! One of the worrying consequences of this infatuation with the law of attraction is that today, people who are truly ill will end up in the hands of "therapists" who think they detain real knowledge. We must highlight that the latter aren't necessarily armed with bad intentions. Many of them have paid a high price for their training, convinced of the scientific legitimacy of the law of attraction because of their illusion of knowledge.

* * *

While we're all constantly flooded with information, the challenge is less to fight against ignorance than against illusion of knowledge. It's easier to teach someone who knows they know nothing than someone who believes they know while they don't actually know.

9

THE IMPORTANCE OF CONTEXT

*"Without context, words and actions
have no meaning at all."*
Gregory Bateson

You probably know the Parable of the Good Samaritan
in the Gospel according to Luke:

*A certain man went down from Jerusalem to Jericho,
and fell among thieves, who stripped him of his rai-
ment, and wounded him, and departed, leaving him
half dead. And by chance there came down a certain
priest that way: and when he saw him, he passed by
on the other side. And likewise a Levite, when he was
at the place, came and looked on him, and passed
by on the other side. But a certain Samaritan, as he
journeyed, came where he was: and when he saw
him, he had compassion on him, and went to him,
and bound up his wounds, pouring in oil and wine,
and set him on his own beast, and brought him to
an inn, and took care of him.*

The distress of others doesn't get to us all in the same way: why did the Samaritan stop to help the man in danger, when the priest and the Levite, though both religious figures, passed him without stopping? Which factors stopped them from doing a good deed? Two millennia later, John Darley and Daniel Batson,[70] both psychology researchers at Princeton University, wanted to answer this question to determine which contextual factors could have an impact on human beings' motivations, their inclinations and thus their actions. For their experiment, Batson and Darley decided to recruit seminarists: nothing better than clergy people to test a parable from the Gospel! The two researchers made the seminarists believe they were going to attend a study on education and religious vocations. Batson and Darley asked each of them to prepare a short presentation on the Parable of the Good Samaritan or on the reason why they took the habit. Then they told them that, because there wasn't enough space in the building, they would have to go to another wing of the campus, (after they prepared their presentation), where they would speak in front of the students. An actor playing the "victim" role was laying on the ground in the courtyard separating the two buildings, so that the participants could not miss seeing him when they would travel from one wing of the campus to the other.

Beforehand, the seminarists had been split into two groups. The ones from the first group had to draft their text as quickly as possible before switching building. The ones from the second group had no time constraint. The results of the experiment show that, as they walked past the courtyard, only 10 per cent of the seminarists of the first group stopped to help the victim, while 63 per cent of the ones from the second group went to aid him. The "time" variable took precedence over everything else. The time pressure seemed stronger than moral values and prevented the first group of seminarists from feeling empathy towards the victim. Darley and Batson noticed that when they pressed some of the seminarists from the first group to hurry up, telling them they were going to be late, these participants actually stepped over the victim to get to the conference room more quickly. The lesson to learn from the Parable of the Good Samaritan is that maybe the priest and the Levite were simply in a hurry.

Context has therefore a big influence on the decisions we make and even on our psychological impulses (empathy, compassion…). The weather,[71] the time of day[72] or our "internal" context (satiety, fatigue, anger, fear…) modulate our way of behaving in the world and towards others. However, if we learn that someone could have helped a person in danger but did not do it, our knee-jerk reaction will not be to tell ourselves: "If this person didn't act, it's because it was raining," but rather: "What a bad person." When it's not about us, our storytelling brain tends to overestimate the responsibility of others and underestimate the context factor.

Yet we also happen not to act the way we should have done, or even to commit shameful acts while trying to shirk our responsibilities. If on the motorway a car cuts in front of another, we rail against them: "What a reckless driver – driving like this is so dangerous and irresponsible!" But if we're late for work and we overtake someone the same way, we will tend to think: "It's not that bad, I'm in a hurry, it's just this once…" We judge others on their actions, but we judge ourselves more leniently on our intentions, because we have access to them.

These double standards between judging ourselves and others is the correspondence bias at the heart of all our social interactions. It underpins many of the other biases previously mentioned. Identified by Stanford professor Lee Ross, this form of double standard judgement creates an imbalance in the way we interact:[73] whether we invoke someone else's responsibility ("What a coward, he didn't come to the victim's aid") or the context when it's about us ("It's not really my fault, it was raining"), we constantly look to accuse the other and to evade our own responsibility. We have this natural tendency to believe that everything should always be someone's fault, but never our own.

Default Choice

External factors often influence our choices. In some cases, a default choice seems to have been made for us. We're exposed to it as citizens, often without knowing it, while the social consequences are often enormous. Two psychology researchers, Eric Johnson and Daniel Goldstein, took a special look at organ donation and the reasons why people choose to donate or not. From one country to the next, the differences in organ donor rates are in fact huge:[74] in Denmark, only 4.25 per cent of residents are donors, while 85.9 per cent of Swedes are donors. And yet these two countries are culturally and sociologically close. It's the same for Germany (12 per cent organ donors) and Austria (99.98 per cent), while these two countries are also quite close on political and societal levels. This gap is explained by the default choice made by each state. Germans and Danes aren't automatically organ donors, they have to actively sign up to a list. By contrast, Swedes and Austrians are organ donors "by default", and must take active steps if they don't want to be. Many Swedes and Austrians haven't taken the step to opt out of these lists. And many Germans and Danes haven't taken the step to opt in. The same inertia on both sides then, but with opposite social and medical consequences: in January 2018, the number of organ donors in Germany had reached its all-time low, and German patients waiting for transplants had become dependent on other EU[75] countries. For all that though, can we judge German citizens as responsible for this crisis? Are Swedes and Austrians more altruistic? No. The administrative context accounts for the percentage of donors and non-donors in those countries, not each citizen's individual will. If you ask a German or a Swede if they're an organ donor, they will probably reply that they have no idea: we aren't often informed of choices made for us, and not only in the context of a state's default choices.

Many websites pre-tick boxes in forms they send to their users. This is how we sometimes find ourselves subscribed to travel insurance policies without knowing, receiving ink-cartridge newsletters when we don't own a printer, or paying for an Amazon Prime subscription while we thought we'd signed up for a one-month free trial. With an increase in these questionable practices, a European law was voted in 2014 to condemn automatic online pre-ticking. In France, the Hamon Law was passed to strengthen these measures. This did not stop sites from continuing, exposing them to heavy sanctions: as recently as 22nd January 2019, the French agency for data protection, CNIL, fined Google 50 million euros for a breach of trust toward its users. Google exploits users' data without asking for their consent explicitly: to the question "Do you allow

Google to use your personal data?", the "I consent" option is pre-ticked, after being drowned in a mass of terms and conditions which are often jargonistic and illegible.

Default choice is everywhere without us being aware of it: if we want to know if someone is writing to us from their mobile phone or from their computer, we just need to look at the first letter of each new message they're sending us. If they're writing from their mobile phone, the first letter will more likely be in upper case, because most smartphones autocorrect to upper case letters by default at the beginning of a sentence. On the contrary, if they're writing from their computer, the first letter will more likely be lower case. All the way down to small everyday things, we are not always facing a free and conscious choice.

Nudges: when you're being whispered the right decision

The upsurge of default choice in our lives pre-dates the internet. In the 1990s, Aad Kieboom and Jos Van Bedaf, two janitors at Amsterdam-Schiphol Airport, noticed that men struggled to aim when they urinated, which significantly increased their workload. One day, they placed a fly sticker on one of the urinals, wondering if the users would attempt to aim at the fly. It's indeed what happened. Following this, flies were engraved on to the airport urinals and, in just a few months, Schiphol's cleaning expenses were lowered by 80 per cent. Influencing someone's choice without them being aware of it (but without causing them harm) can therefore be in the common good. This method is called a *nudge*, and in this context, is a psychological prompting that gently influences human behaviour, supposedly for our own benefit.

Road safety signage sometimes uses *nudges* to incite drivers to adopt a more responsible behaviour on the road. Chicago has a dangerous turn on a cliff road with a breathtaking view. In front of an increase of road accidents there, the city council decided to paint the white stripes on the road closer and closer as they get nearer to the turn. This gives the drivers an impression of speed, which pushes them to slow down. This simple optical illusion lowered accidents by 36 per cent in this location. The same mechanism is at work for pedestrian crossings in 3D since the 26 of June 2018 in some streets of Paris's 14th *arrondissement* , giving vehicle drivers the impression of a physical obstacle, pushing them to slow down more than with a traditional pedestrian crossing.[76]

Other nudges use default choice. To encourage us to be more respectful of the environment, we're no longer given plastic bags by default in supermarkets. In the United Kingdom, the government's Nudge

Unit modified the British organ donation webpage by adding this sentence: "Every day thousands of people who see this page decide to register", betting on a desire to conform socially: if thousands of people register, we will want to do it too. Within one year of this, the registration rate went from 2.3 per cent to 3.2 per cent, i.e. 96,000 additional registrations.[77] In France, "gentle prompts" are gaining ground: a behavioural science public policy unit has been set up by the DITP, a cross-ministerial working group devoted to the modernization of public life. Following this, in Lyon, the metro's stairs were coloured to encourage passengers to prefer them over escalators, and therefore to do more physical exercise. In Nantes, sorting bins were painted by artists to raise the residents' awareness of the necessity to sort their household rubbish.

Nudges are based on cheap and simple levers, which is why public authorities are taking an interest in them today. It is a way to convert citizens to more civic-minded behaviour, and at a low cost. It must nonetheless be highlighted that it has been proven that the effect of nudges diminishes and even disappears when people become aware of them or get used to them.[78]

The influence of social context

The ways we behave are of course influenced by others, even if this is often unconscious: hypersensitivity to social signals shapes and conditions us. To live together, share values and moments of togetherness, to organize ourselves in groups, cities, states, we need to be aware of the signals other people send. Our musical or culinary taste, our religious beliefs, our childhood friends, the colours we like, the football team we support, the political party we belong to and so many more of our personality traits largely depend on our place of birth, on the town we grew up in, of our entourage's social condition. This is what we call social context: the factors we didn't choose and which lead us to have such tastes, such certainties and such beliefs. In other words, our culture.

Social Compliance

There are group dynamics that sometimes push us to act out of mimicry and not out of conviction or true desire: social mimicry. I'd like to tell you a personal anecdote about this: when I was child, we used to play a game at school. We were about thirty-one pupils* per class. About ten of us used to agree to all look at the ceiling at the same time, all at once. The other pupils,

---

* You might feel like thirty-one is a precise number, while had I said: "about thirty" you would have felt that it was a more *plausible* approximation. However, it's a familiarity bias, and thirty-one is as approximative an estimate as thirty or twenty-five. Indeed, to say "I'll be here in thirteen minutes" isn't more specific than to say "I'll be there in ten minutes", even if this sentence gives this impression because we have a tendency to use "round" numbers more often in our approximations.

who didn't know about our stratagem, mimicked us and also gazed up at the ceiling. There was absolutely nothing there, but it was so funny to see everyone (even our teacher) fall for it! A social signal that's been coordinated in advance affects the behaviour of others because the need to mimic is deeply anchored within us; it is manifested as early as childhood.

Social psychology pioneer Solomon Asch ran an experiment in 1951[79] which tended to prove that even in the context of a simple perceptive task (meaning the evaluation of what's before our eyes), group dynamics influence individuals who are part of it. For this, he gathered a few of his students and presented them a card marked with a line, then a card with three lines of noticeably different sizes, as you will see below.

He asked them to identify which of the three lines on the second card matched the figure on the first one. The task is easy to carry out (this is intended), and each person had to give, each in turn, their answer in front of the group. All the students are in on it, except for one, the "real" guinea pig. The "fake" participants have been instructed to give the same answer each time, whether it's true or false. On the eighteen tries that Asch made them do, the "fake" participants were instructed to give a wrong answer twelve times. The results show that the "real" guinea pig also got it wrong in almost 40 per cent of cases, which is huge for such an easy task. To have something to compare this with, Asch had given the same exercise to a "reference" group of students: these had obtained 99 per cent correct answers. Peer pressure easily drives us to making mistakes on elementary exercises.

You've probably already heard the following piece of criticism: "People are sheep, they want others to think for them…" It is generally expressed by people belonging to a certain social category, where this opinion is fairly widespread. This criticism of compliance is therefore, without the speaker realizing it, an example of this much-maligned compliance!

The weight of social conformism was brought to light by one of the most famous social psychologists of the twentieth century: Stanley Milgram. To determine the extent to which human beings may lose their judgement faculties when facing an authority figure which has been recognized as legitimate, Milgram gathered a group of men and women. Facing them was a man, sat on a chair, electrodes on his body (an actor, but the participants didn't know this). After they were given a remote control that triggers electric shocks, Milgram asked the participants to electrocute the actor, who faked extreme pain, increasingly so after each discharge. Milgram explained then that the number of volts increased throughout the experiment. In most cases, the participants continued to press on the remote control even when the number of volts was high enough to lead to death. This experiment was reproduced in a television show called *Le Jeu de la mort* (*The Death Game*) broadcast on French channel France 2 in March 2010. This time, the participants are greeted by an audience in hysterics, and win money with each discharge sent. Encouraged by the presenter (the authority figure), surrounded by a cheering audience and enticed by financial gain to boot, the participants continue to send electric shocks, not because they're sadistic or inhuman, but because the context encourages them to do it; 81 per cent of the participants would have caused the death of the man in front of them.

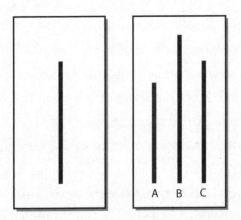

We use the example of Milgram and of experiments stemming from his because they're fairly well known and reveal that situations have a worrying effect on human behaviour, but it's necessary to know that these results are quite controversial among psychologists and that his vision of obedience to authority is a bit of a caricature. People tend to resist authority more than Milgram lets us think. Recent studies show that most of the guinea pigs seemed to know deep inside that this was an experiment and that the man in front of them wasn't really in danger of death.[80]

We can indeed be blinded by our unconscious desire to socially conform, but in some situations, conforming to the choice of a group is rational and even advantageous from an evolutionary standpoint. It's the case for example of the contagious stress mentioned in Chapter 4. Stress is an easily communicated social signal, which enables us to react to danger even if the latter isn't directly visible to us. If I see a crowd of people running in the street, without understanding why, it's more advantageous for me to start running in the same direction than not to run and thus run the risk of facing the danger the others fled.

Beyond the safety aspect, social compliance also creates cohesion between human beings and people. Phenomena like the Korean pop dance routine 'Gangnam Style' in 2012, or the 'Macarena' (which never went out of style) ran riot across the world. All over the world, everyone dances to these rhythms in the same way, which contributes to reinforce the sentiment that we share the same world. This idea is the basis of the pop culture principle. Great sporting events also bring people together: The French national team's World Cup victory on 15th July 2018 created a great wave of national cohesion in France. At the very beginning of that same World Cup, when Mexico scored a goal against Germany, Mexican seismographs even detected an artificial "earthquake" triggered by the Mexican fans' joyful jumps![81]

Interested in the matter of crowd wisdom, the Argentinian researcher Joaquín Navajas[82] asked a sample group of 5,180 people to reply to a list of general knowledge questions: what's the height of the Eiffel Tower? The number of people in Mexico? The surface area of Brazil?... The experiment took place in three steps. They first had to answer individually. They were then asked to gather in groups of five and to talk through them for a few minutes to reach a consensus. Finally, they were all given the option to go back on their original answer. The answers generated by the small groups of five were significantly more accurate than the combined average of the individual answers. The number of people making up a decisional group is therefore an important factor in the relevance of the results. The more we are, the less our answers are precise. But if we form small thinking groups, we will be more pertinent than if we think alone.

Social context weighs on our decisions, above all when this social pressure isn't identified. You may already have noticed cups and balls tricksters when walking in a touristy area of a large city. When you walk past a cups and balls stall, you see a dealer playing with their cups and balls alone. Intrigued passers-by look at the stall, occasionally stopping to watch or play. Yet three or four of these are in fact accomplices. Their work is to create a sentiment of social compliance, which will make you want to stop and play yourself. You will spend a bit of time watching the game, see that people around you (the accomplices) win considerable sums easily. You tell yourself that it seems easy and that you could also play. Then the dealer will start to play with you, to create a feigned complicity with you; they will for example show you where the ball is in too obvious a way while one of their accomplices is playing and loses on purpose to trigger an emotional reaction in you, which in turn launches automatic thinking such as: "Are they stupid? It's obvious that the ball is under this cup!" Seeing that you're biting, the dealer will entice you with a free go, which they'll let you win; then they'll make you enter dissonance by letting you know you could have won 200 euros if you had bet money... To reduce the dissonance, you'll therefore have to bet, but you have practically no chance of winning: the setup is rigged and the dealer has dozens of ways to make you lose. In no time, you will find yourself with an empty wallet. If one day you stumble upon a cups and balls table, take a good look at it – it's fascinating when we decode the operation – but don't play thinking you can win: you're guaranteed to lose! When we interact with others, under the gaze of others, we are made to feel an unconscious social pressure which can sometimes push us to behave against all common sense.

Group Behaviour and (In)action

A patient once told me of a traumatic experience which had happened to her when she was taking the bus in the middle of the day in Paris. The bus was half full, but the back of the bus was empty, so she sat there. Three men got on and surrounded her. They first teased her by making a few unsavoury jokes, then they became more physically aggressive, and this in full view of the other passengers. The young woman first raised her voice and said "Don't touch me", then one of the men snatched her tote bag, yelling "Tell me what you have in this bag", while another man was ostentatiously stroking her thigh telling her, "You're scared, huh?" Terrorized, the young woman fled at the next stop. When the doors closed again, she heard them scream, "Whore!" She then told me what struck her the most: "If I take the bus rather than the metro, it's because I know that in the metro nobody would lift their little finger if something happened to me. In the bus, there's a driver, people are closer, and yet, all I could observe was an awkward indifference from all the passengers when they'd all seen that I was getting harassed."

Listening to this story, many of us could be led to thinking: "Had I been in their place, I would have done something." And it may be true. Only, there are strong chances that your actual behaviour might not have depended on the gravity of the situation as much as the number of people there when the attack took place. Two American psychologists, John Darley and Bibb Latané, demonstrated that there was a correlation between the number of people present at the time of an accident or an attack and the action or inaction of the witnesses at the scene. After the simulation experiment of an epileptic seizure in a public place, they noted[83] that 85 per cent of passers-by helped the victim if they were alone at the scene of the incident, 62 per cent if there was someone else, and only 31 per cent if there were more than four other witnesses. The more witnesses, the less people act, because they don't feel as responsible than if they had been alone and that everything depended on them. This mechanism is called the bystander effect: I think that someone else will act, but that someone else will think the same, so in the end nobody acts. In Paris, a poster campaign was launched on public transport for each and every one to regain awareness of their individual responsibility: in March 2018, the RATP launched an anti-harassment billboard campaign on Parisian buses and metros, and since then regularly broadcasts the message: "If you're the victim of witness of harassment, send us a text message to 31 177 or dial 31 17."

Solidarity Chains

These group dynamics can lead to cowardice but also to solidarity. Indeed, it sometimes only takes the action of one person to create collective momentum. In 2015, in Lebanon, a "rubbish crisis" lasted several months without anyone reacting. One day, a small group made up of about fifteen young people who had had enough of it set up in front of the parliament with signs and banners that expressed this exact sentiment. They were soon joined by about a hundred people, and in the end several thousands of protesters were in the streets. For a country of four million inhabitants, this was one of the largest independent people's movements in the history of the country, and it started with the initiative of a very small group of citizens.

It is the same mechanism we find at the origin of the great uprisings which have marked history. One of the most recent was the one we called the "Arab Spring"[84]. In December 2010, Mohamed Bouazizi, a twenty-six-year-old Tunisian street vendor, gets his merchandise confiscated by policemen who beat him to a pulp. One more time. One time too many. From a very poor family, Mohamed had to quit high school during his final year to support his brothers and sisters financially. After several years of unemployment, and given the employment shortage in Tunisia, he accepts to sell fruits and vegetables illegally. The victim of numerous humiliations from the police, without the means to pay the fines, he sets himself on fire in front of the governor's office on 17th December 2010. A large part of the Tunisian population can no longer stand daily bullying by the police, who are a symbol of the dictatorship of President Ben Ali, in power for twenty-three years. This isolated case of hopelessness and protest becomes the trigger of a national revolt. A few days later, thousands of Tunisians walk behind his coffin. A few weeks later, President Ben Ali flees the country and the Tunisian regime collapses. Revolts in Arab countries later reach Egypt, then Libya and Syria.

On a smaller scale, we can also cite the Gilet Jaune ("yellow vest") movement in France. On 10th October 2018, Éric Drouet, a lorry driver, creates a Facebook event to call for a protest against a rise in fuel prices announced by the government. Three days later, Priscilla Ludosky starts an online petition also demanding lower fuel prices. Very soon, hundreds then thousands of people (1.1 million in one month) sign it, and tens of thousands of people take to the streets, occupy roundabouts and let their frustration out. The movement even crosses French borders, and the yellow vest becomes a symbol of protest and anger against pauperization and the decrease in spending power.

10

A TOOLBOX FOR MORE MENTAL FLEXIBILITY

"You know what thinking is? It's just a fancy word for changing your mind."
Doctor Who

There isn't a magical spell to undo the traps our brains sets for us. No simple and immediate solution to "unbias" us. But we can try to counter the biases' negative effects by taking an interest in the mechanisms that trigger them. Therefore, here is some practical advice to react when our brain works against us.*

Beyond automatic thinking

The human brain naturally produces thoughts, emotions and automatic actions, which psychologists call heuristics, and which enable us to evolve in a world that's too complex to be apprehended with all its nuances. Added to these heuristics are secondary thoughts, thoughts about our thoughts, expressed by a little voice in our head: we call them metacognitions.

---

* Updates on these tips and tricks will be regularly offered on the webpage of the chiasma.co collective as this research progresses.

It's on these we will be able to act in situations where we gather that we're victim to certain cognitive biases causing us harm.

Let's imagine the following situation: Victor is a jealous man and his jealousy is wrecking his life, provoking stress reactions as soon as he thinks his girlfriend is unfaithful. When he calls her up and she doesn't reply immediately, his first thought is to tell himself she's cheating on him. An inner voice whispers to him: "She's cheating on you, it's certain." This thought about his thought reinforces his primary thought and urges him to call his girlfriend more and more often. Each unanswered call will be a new proof of infidelity. In this case, metacognitions reinforce primary thoughts, making the situation even more problematic: it's a vicious circle.

We have no immediate control over our primary thoughts, which are too quick and automatic, but it's possible to act on metacognitions. This metacognitive control aims to delegitimize our harmful automatic thoughts. We don't have to feel responsibility for or ownership of our primary thoughts: nobody chooses to be jealous, stingy or petty. But we can and must act on our metacognitions. It's why we must learn to identify the primary and automatic thoughts and emotions linked to a problematic situation, then to create a distance between the primary thought and the metacognitions. The latter can no longer reinforce the primary thought. Imagine that Victor has undergone metacognitive training linked to jealousy situations. He calls his girlfriend and she doesn't pick up. His automatic thinking is triggered by his little inner voice ("she's definitely cheating on me"). Instead of letting it speak up, he will doubt this little voice by confronting it with another inner monologue: "she might be on the metro, at a friend's or talking to her mother, and she'll call me back later." This will calm the physiological stress reaction linked to the idea of jealousy. Victor will be able to breathe more peacefully and it will be less difficult to leave his phone alone and to get back to his activities. Victor has reduced the influence of his primary thought. By repeating these metacognitive control exercises, Victor has gained mental flexibility and will have more chances to heal from his crippling jealousy.

These metacognitive control techniques are increasingly utilized as therapeutic tools in clinical psychology.[85] I use them myself with my patients. We can add other methods to make them even more effective. Fairly similar to metacognitive control techniques, psychoeducation gives transparency to the psychological mechanisms at play in the patient's brain. The aim of psychoeducation techniques is to help people suffering from psychological illnesses develop their own solutions by better understanding the psychological cogs at the root of their phobias and anxieties. I've recently had a consultation with a patient who suffered from Obsessive Compulsive Disorder (OCD). For him, everything that isn't practically perfect leads to suffering. He explained to me that he noticed things a healthy person wouldn't even see (for example, an object that isn't exactly in its place, a small bit of limescale on a glass...) and this triggered terrible anxiety in him, even suicidal impulses. So I explained to my patient that he was suffering from a completely binary view of the world: for him, everything that isn't perfect is absolutely rubbish. I then recommended that, each time he felt an episode approaching, he always kept in mind that his anxiety came from this binary thinking. Even if this technique alone will not enable him to heal from his obsessive disorder, the aim is to initiate an improvement in his way of handling the thoughts that provoke his anxiety.

If you suffer from a psychological pathology like OCD or social anxiety, or even negative thoughts and emotions, the idea is to put the metacognitions coming from it to three fundamental questions:

- What concrete elements form the basis of your automatic thought?
- Is this thought or emotion sterile and circular?

Does it come back periodically to lock you up inside a vicious circle?

- What would you recommend to a friend if they shared a thought like this one with you?

This questioning will progressively enable you to gain some distance each time anxiety-generating situations come your way, and will eventually enable you to limit the appearance of harmful automatic thoughts.

Keeping the extent of our knowledge in check

Operating with approximations and shortcuts, our brain can expose us to mistakes, especially since we often have incomplete information and very limited knowledge on most subjects. However, we like to believe that our beliefs and opinions are justified, because being wrong is unpleasant.

To determine if we are right to be sure of ourselves, we will therefore focus on the reasons why we have a given opinion, rather than on the opinion itself. This will enable us to re-evaluate our opinions more easily according to new information we receive, rather than staying attached to the opinion at hand. We can also allocate a trust index to our beliefs and our opinions – those we have had for a long time as well as those we've recently acquired. For this, we will allocate a reliability percentage (more or less high) for the knowledge we have on a given subject. The aim of these trust indexes is to know at which moment to doubt and at which moment to trust ourselves by adopting a mode of thinking that is progressive (I know a lot about/I know little about this) rather than binary (I know/I don't know).

In his book *The Demon-Haunted World*,[86] astronomer Carl Sagan set out a very well-stocked toolbox to evaluate the level of reliability of the information we receive and that we're susceptible to believe. My Chiasma Association colleagues and I drew a lot of inspiration from it to create a survival guide in times of elections (i.e. when the use of rhetoric is the most frequent). Obviously, the same advice is also valuable outside election times.

Recommendation 1: be aware of **ad hominem arguments,** meaning an attack aimed against a someone only because of their position or status. Far-right French presidential candidate Marine Le Pen, during her 2017 debate against her second-round opponent Emmanuel Macron, was an *ad hominem* champion: before attacking her rival on his electoral programme, the candidate accused him of being a "beloved child of the system and of the elites", "Hollande Jr", "the candidate of unhinged globalization", etc. Donald Trump regularly uses the same technique: in 2016, on

Twitter, his argument against Mitt Romney, one of his most serious critics within his own party, was that he walked like a big dumb bird!

Recommendation 2: Watch out for **authority arguments.** When someone uses their rank, position or profession to make a point, we need to examine that argument's validity as if it had been brought forward by an average person. At the end of his first mandate in 1970, Richard Nixon won the favour of American voters' by promising he knew how to put an end to the highly unpopular Vietnam War, without explaining how he intended to do this. He used his status as president to convince Americans to vote for him, while the surge in American bombing on Vietnam contradicted all his campaign promises.

Recommendation 3: spot **false analogies**, arguments which rest on a parallel between two things or two situations that have so little in common that comparing them is unjustified. For example, during the 2017 French presidential elections, far-left first-round candidate Jean-Luc Mélenchon declared, for example: "I am not for different system of labour laws for each company, as I am not for different highway codes for each street." As attractive as this slogan might be for some, we can't forget that fundamentally there's hardly anything in common between labour laws and the highway code.

Recommendation 4: do not give in to **calls for an emotional response**. For example to say that today, in France, we open our borders left, right and centre, that we want to welcome all the world's misery while millions of French people are unemployed or live below the poverty threshold… the aim of this argument is to incite a strong emotion (often fear or anger), to trigger a visceral reaction leading to the adherence to certain ideas that won't have real factual evidence behind them.

Recommendation 5: favour scientific proof over **anecdotal evidence**, which consists in drawing a general conclusion from an isolated fact or example. For example: "We must ban war video games, because a student attacked one of his classmates after playing one." This statement doesn't take into account all the other teenagers who have played the same game and didn't end up pointing a gun at their peers. Furthermore, this statement doesn't clearly explain the link of cause and effect between playing a violent video game and attacking fellow teenagers.

Recommendation 6: avoid **false equivalences**. The fact that a politician has been guilty of misdeeds doesn't mean we can state that all political authority figures are "rotten". Nuance is in order.

Using these tools to face misinformation

The late Umberto Eco wrote in one of his columns that, in the past, a fallacious idea died when there was nobody left to listen to it any more, while today it can spread like wildfire and reach the greatest numbers. Cambridge University Professor Sinan Aral[87] added numbers to this idea: in an article published in 2018 in *Science Magazine*, he reported that MIT tracked the internet circulation of 128,000 pieces of information, half false, half true. Analysing the results of this study, he noticed that fake news travels six times faster than real news: simplicity, extravagant language, disgust and surprise are the primary fuels that generate clicks.

In India, in August 2017, riots broke out in the Banda region following a message sent through WhatsApp: a fourteen-year-old girl had been attacked by a guard in the local market of Gurdwara. This piece of news, having gone viral in two days, led to violent confrontations between the Sikh and Hindu communities. Violent actions which could have been prevented because the information that triggered them was false. Still in India, seven men were lynched and beaten to death in the State of Jharkhand in May 2017[88] following a fake child-kidnapping rumour spread on WhatsApp. Two months beforehand, two men wrongfully suspected of shoplifting on social media had suffered the same fate. In response to this tsunami of violence, the Indian government set up internet blackouts, sometimes partial (only affecting chat apps), sometime complete. Between January 2016 and May 2018, with 154 blackouts, India was the country which, on the global scale, cut off the internet the most frequently, far ahead of more authoritarian regimes like Pakistan (nineteen blackouts), Iraq (eight blackouts) or Syria (eight blackouts).

Fake news spreads like a virus during an epidemic. British researcher Gordon Pennycook[89] wanted to find solutions on the individual scale to avoid contamination and involuntary propagation as much as possible. Pennycook gathered a panel of internet users, to whom he showed a series of tweets and articles on social and political matters, which included some fake news. He split the participants in two groups. To the first group he gave no instructions other than determining which where the truthful pieces of information. A computer was at their disposal, but he did not explicitly recommend they use it. But he asked the participants of the second group to have a good think and take a step back before answering, and pointed out he had put a computer at their disposal to help them check the validity of the statements. The idea was to incite this second group to implement a form of metacognitive control which would prevent them from trusting their primal and automatic thoughts.

Results show that many more participants spotted the fake news among the second group. Adopting an analytical reasoning by putting a metacognitive control in place enabled the participants of the second group to see through fake news and clever formula aiming to give the appearance of truth to false information. The first group, on the other hand, fell victim to more fake news, even though it wasn't made of politically engaged people. Pennycook called this group the "lazy, not biased" group. In a world where each of us has quick and simple access to a profusion of information, this laziness exposes us to making mistakes. In order to be less of a victim, we have to be less lazy, less passive, more doubtful and desirous to check things through reliable sources.[90]

Stanford University researchers Sam Wineburg and Sarah McGrew[91] gathered average internet users and professional fact-checkers to show there is a right and a wrong way to read a web page. The two researchers asked all the participants to read the same page, then to tell them if they had thought that the information contained was true or false. They observed that the uninformed had a tendency to read the page from top to bottom, without checking the validity of what was in front of them, while the fact-checkers systematically opened several tabs as soon as a piece of information seemed suspicious. At the end of the experiment, the latter had spotted practically all of the fake news, while the "uninformed" users had almost spotted none. We should therefore always prefer a horizontal read of the web over a vertical one, in other words, not remain passive in front of a page, but open others.

When Google and Facebook fight against misinformation

Fact-checking work is lengthy and expensive, and faced with the masses of information circulating, traditional media fact-checkers soon transform into Sisyphus, pushing their rock towards the top of the hill each and every day. A rock that doesn't cease to roll back down under the weight of new information to fact-check. Claiming to want to help these fact-checkers with their work, Google, WhatsApp, Twitter and Facebook[92] developed a "trust indicator" which would take the form of an icon indicating the highest or lowest reliability of an information source. The project bears the name Trust Project. Yet these social media platforms are the main location for fake news propagation. Can the principal vector for contamination be the remedy?

We could think that they are not to be held accountable for this propagation because they are only broadcasters. In the same way that a pipeline company isn't responsible for the quality of the water circulating through its infrastructures, Facebook or Twitter couldn't be held accountable for misinformation circulating through their networks. Except that by editorializing the content of our searches, i.e. by analysing our personal data to then present us with content in line with our beliefs and matching our interests, Google and Facebook partake in the creation of this content and aren't mere "pipelines" moving millions of pieces of information without tampering with them, but pipelines which modify the fluid that comes out of each tap.

In March 2018, the *Guardian*, the *Observer* and the *New York Times* revealed that Cambridge Analytica, a private internet data analysis company had harvested millions of personal data from Facebook users via common Facebook quizzes (e.g. "Which *Harry Potter* character are you?", "What would have been your given name in prehistoric times?", etc.) The questionnaire didn't just absorb the participants' data, but also their Facebook friends'. And yet, before this scandal burst out, Facebook had always advocated a transparency policy towards its users, promising since 2011 to notify them as soon as their data might be used and distributed. British Channel 4 broadcast a report filmed inside the headquarters of Cambridge Analytica, during which Alexander Nix, the company's CEO, confirmed that these practices extended to the wilful spreading of misinformation, spying on political opponents and using individuals' personal data to manipulate public opinion.

Furthermore, when public authorities set up policies to fight against misinformation on a national scale, these tend to limit the freedom of the press. In France, Emmanuel Macron shared his intention to create a regulatory body... targeting journalists, as well as a "deontological council of the press" whose outline is still obscure and which could be a real danger for the freedom of the press. The place where the state should play a role is in the fight for more transparency from Twitter and Facebook, the fight to trace online personal data and above all to enable each citizen to have access to their personal data, control how it is used and be able to delete it freely.

In the meantime, we have no other choice than to individually DIY our own information consumer hygiene by trying to apply the trust indicator technique as much as we can; this requires a learning phase which could take a long time. In psychoeducation therapy, we must work for several months with our adult patients before they manage to evaluate their trust indicator for their opinions and their beliefs. This process would be quicker if critical reasoning were taught as early as possible. Researchers produced handbooks for achieving a more critical mindset, i.e. an intellectual attitude capable of adapting when facing any information, which examines it attentively through reason, gathers material about it and submits it to scrutiny before believing it to be true or real.[93] From elementary school until the end of high school, classes should be structured around several themes: how to conduct observations and interpret them; how to determine the causes and effects of various events or mechanisms; how to evaluate the reliability of information sources and content relayed in texts, images and videos; finally, how to debate scientific or social subjects. This thorough training of our critical reasoning techniques is one of the rare tools to be unanimously approved by cognitive science researchers in the struggle against misinformation.[94]

Conclusion

REGROUNDING OURSELVES
IN A SHARED REALITY

Biased thinking has always existed, there has always been a left and a right, bossy and permissive parents, open relationships and traditional ones, hard workers and idlers, optimists and pessimists... But before our digital era, political, or social stances, even if they were at odds with each other, all co-existed within the same reality. We used to share the same world beyond our disagreements on the way live in it. Today, the situation is different: when Donald Trump declared in a November 2018 tweet: "Brutal and Extended Cold Blast could shatter ALL RECORDS. Whatever happened to Global Warming?", he gets weather and climate mixed up. This tweet was liked by thousands of people, who thus adhered to "alternate facts" – an oxymoron coined by the Trump team – in which global warming doesn't exist. The world we live in has become a world where known facts and false beliefs are on an equal footing.

The more we'll stay confined "among ourselves" through social media, the more our identity will take the colour of the group we frequent, and we will be swayed to reject any conflicting voice. This mechanism leads to a ghettoization of society, which risks tearing our social fabric to shreds. This begs the question with increasing urgency: "What do we have in common?" The first thing we have in common is the world. The reality of facts. Global warming is as real as the desk at which I'm writing. Men have really walked on the moon. The earth is round. To deny these facts would be to deprive ourselves of any common ground, and make any civic life impossible and the world an unbearable and inhabitable place. Now that social dialogue is turning into a series of monologues, that opinions are becoming radicalized and society polarized, that violence grows, we must preserve and develop our shared reality in order to rebuild a space for democracy. The best tool to do so is through doubt: a constructive doubt, turned towards oneself, and not an incriminating doubt turned towards the others.

But doubt is a double-edged sword that one must learn to handle cautiously. This is why it is necessary to adopt nuanced rather than absolute doubt. A progressive doubt that we have to apply on a case-by-case basis to be able to face the complexity of the world. French mathematician and philosopher Henri Poincaré said: "To doubt everything or to doubt nothing are two equally convenient solutions, which both exempt us from thinking." What I encourage you to do is to cultivate critical reasoning in order to determine when it's good to doubt, and when you can be confident.

Think about the cerebral mechanisms at work whenever you think, whenever you believe, whenever you judge. If you feel your body tensing up because a topic is stressing you out, doubt a little; if you feel a belief is so dear to your heart that you can't bear to see it questioned, you'll know you're partly blinded by motivated reasoning, so doubt a little; if you spontaneously judge someone, ask yourself what your judgement is based on, think about the context again: doubt a little. Remember that this person operates according to the same mechanisms as you, and try to hold off your judgement until you understand what has possibly pushed them to act a certain way.

Knowing how to doubt our thoughts, our emotions and our intuitions when it is necessary leads us to see the world in all of its nuances and its complexity again, and to free ourselves from tunnel vision. By taking a step back from our convictions, by saving ourselves from adopting a black-and-white vision of people and situations, we offer ourselves a chance to reconnect. Let's all accept to make this effort, so we can mend our social fabric together, resume dialogue, and once again share the same world.

Acknowledgements

Mariam Chammat, without whom this book wouldn't exist, because it is the fruit of more than ten years of exchanges, discussions and debates. My words could have been hers; we are, so to speak, two bodies sharing one brain.

Joëlle, without whom I wouldn't be who I am.

Guillaume Allary, who trusted me with my first attempt. I will always be grateful for the chance I had to have such a kind publisher who knew how to get the best out of what I could create. It's an honour for me to be part of his publishing house.

Louise Giovannangeli, my editor, the best. If I can be proud of this work, it's in great part thanks to what she made out of it; I couldn't do justice to the quality she brought to the text. It was a pleasure working with her and all the Allary Éditions team!

Marc Pondruel for the time he took in accompanying me during long nights of work and guitar-playing.

Camille Rozier and Sami Abboud, my Chiasma acolytes. It's there that the desire to write this book was born.

Thibaud Griessinger for his expertise on the scientific content of the book. Our conversations didn't only enrich this work, but they also continue to enhance my thought process.

Antoine Pélissolo, who taught me to doubt the tricks my brain plays on me and to rise above my knowledge illusions. I will always be grateful to him for this.

My friends. I work to spend more time with them. In alphabetical order: Dima, Hicham, Wiss, Grace, Inkling, Catarina, Chadi, Luca, Mag, Karim, Manou, Samer, Val, Emileo, Juan, Link, Hass, Réa, Paulo, Marco, Lucie, Philippe, Emma, Nadimo, Anwyn, Camil, Edo, Teller, Rachlu, Reddit, Marie-Sarah, Charlotte, Manitou, Chewich, Layale, Akram, Tetris, Laurence, Dan, Jado, Nadim, Hugues, Louise, Marwa, Hervé, Claude, Zelda, Sami, Benjamin, Célina, Guitta, Mario, Chatta, Nana, Nicky, Leia, Farah, Switchy, Stepal.

All the people with whom I was able to interact and who contributed to the tricks my brain plays on me.

And of course my family in Lebanon: Selim, Yvonne, Sissa and Adou for being here since the beginning, quite literally so.

Thank you.

Glossary

A Priori Knowledge: knowledge we have within us, independently from our sensory experience. This can have an impact on the way we reduce the ambiguity of a situation.

Ambiguity Reduction: an often unconscious and spontaneous action which enables us to stabilize an ambiguous image or situation in order to develop a coherent world view.

Analytical Reasoning: neutral processing of all information on a given subject at our disposal, without preconceptions regarding the conclusions this reasoning will lead us to.

Anchoring Bias: tendency to only retain one piece of information to judge a given situation (usually the first piece of information provided).

Anecdotal Evidence Fallacy: tendency to consider an anecdote or an isolated piece of information as sufficient evidence to draw general conclusions on issues that are often complex.

Argument from Authority (*argumentum ab auctoritate*, also called an appeal to authority, or argumentum ad verecundiam): using one's rank, position or profession to convince an audience.

Automatic Thoughts: thoughts that come up too quickly to be directly monitored.

Binary Reasoning Fallacy: tendency to reduce a situation to a black-and-white alternative while the world is complex, more often grey than black or white. Binary reasoning works as a switch which would only have an "on" and an "off" function, while an objective reasoning must function like a dimmer switch.

Bistable Image: an ambiguous image which can only be interpreted in two different ways. There are also tristable and multistable images.

Bystander Effect: psycho-sociological phenomenon making us less responsible of our actions or our inaction in a crisis situation where other people are present and able with the ability to act.

Choice Blindness: a brain function which prevents us from remembering our choices, while we can justify them after the fact if we're asked to.

Cognitive Biases: deviations made by our brain to make decisions or judgements in a less laborious manner than it would through analytical reasoning, which take all relevant information at our disposal into account. Quick and useful, cognitive biases can be at the origin of judgement errors. The list of cognitive biases doesn't cease to grow; here are the main ones: Anchoring Bias, Confirmation Bias, Halo Effect (Notoriety Bias), Representativeness Bias, Selection Bias, Negative Stereotyping Bias, Negative Interpretation Bias, Overconfidence Bias, Present Bias and Fundamental Attribution Error (Correspondence Bias).

Cognitive Dissonance: a feeling of mental discomfort when we harbour thoughts or opinions in contradiction with our behaviours.

Cognitive Homeostasis: state of equilibrium and psychological stability which we spontaneously look for.

Confabulation: imaginary narrative constructed by our brain to compensate for a mnemonic deficiency, especially in neurological disorders.

Confidence Index: a practice consisting of attributing various levels of trust in our opinions and thoughts. The idea is to rid ourselves of our binary vision of things (I know/I don't know) to adopt a progressive vision which leaves us with a greater handling margin to change our mind and gain in open-mindedness.

Confirmation Bias: tendency to favour information confirming preconceived ideas, opinions and beliefs, while rejecting those in opposition with them or which contradict them.

Disinformation Effect: information received as an *after-shock* coming to alter the precision of the memory we have of an event and going sometimes to the point of creating false memories.

Dunning-Kruger Effect: also called overconfidence bias. Confidence peak in our abilities which is manifested each time we discover a new topic and that the first acquired knowledge pushes us to believe we master the subject matter.

Exposure Therapy: a technique consisting of exposing a person suffering with anxiety or phobia to the cause of their anxiety or phobia, in order to observe their emotional, behavioural and cognitive reactions and with time, to help them overcome their fear.

Fallacy of the Single Cause: tendency to believe that an event only has one cause, while it's more complex and multifactorial.

False Equivalence Fallacy: an error consisting in drawing a parallel between two things having some commonality, while they're of a different nature.

Fundamental Attribution Error: tendency to judge others on their actions and not on their intentions, and conversely judge ourselves on our intentions and not on our actions.

Halo Effect: tendency to believe that if a person is famous, their opinion is more valuable than someone who is lesser known but who's an expert.

Heuristic: spontaneous action or thought which gives fairly good results in a given situation and which has the advantage of being quasi-instantaneous, even if it's often too approximative.

Illusion of Explanatory Depth: tendency to overestimate our understanding of how objects work, and by extension, how the world functions.

Impostor's Syndrome: a person's tendency to underestimate their real competence and to believe they can never meet expectations.

Inference: intellectual operation at the base of all reasoning which consists in reaching a conclusion starting from a postulate, going through several logical steps.

Learned Helplessness: behaviour developed when we are in an adverse situation or one of repeated failures, which pushes us to believe that we can no longer change things and that we are condemned to suffer what happens to us.

Locus of Control: self-evaluation of our own ability to control what happens to us. If we consider that what happens to us depends on us, we have an *internal* locus of control (ILC). If we consider that our life is ruled by external factors and that we have no control over what happens to us, we have an *external* locus of control (ELC).

Mental Flexibility: ability to change one's mind and to update our beliefs according to the new information we receive and the new experiences we live.

Mental Rigidity: refusal to change one's view point and to update our opinions and beliefs according to new information we receive when they don't go our way (the opposite of "mental flexibility").

Metacognitions: from "meta" (beyond) and "cognitions" (thoughts). Thoughts that come on top of our automatic thoughts. They correspond to a little inner voice which makes itself heard when we think about something.

Metacognitive Control: work we can do on our metacognitions to distance and re-evaluate our automatic thoughts or emotions.

Misinformation: a new term used to talk about fake news, false and unfounded information.

Motivated Reasoning: a way of reasoning which only pays attention to what confirms our beliefs, rejects what calls them into question (see "confirmation bias") and develops justifications to rationalize these beliefs a posteriori, thus comforting us with the idea that we are right to think what we think.

Negative Interpretation Bias: tendency in a situation which can be interpreted negatively or positively, a tendency to reduce the ambiguity by choosing the negative option.

Negative Stereotyping Bias: tendency to believe and spread negative stereotypes about a given category of people.

Nudge: behavioural technique based on simple levers (optical illusion, default choice, social conformity...) in order to incite rather than force individuals to adopt a more civil behaviour.

Object Permanence: awareness that objects external to us continue to exist in space even when we no longer see them. This faculty is shared by human beings and animals.

Overconfidence Bias: tendency to overestimate our capacities or our knowledge on a given subject.

Present Bias: tendency to grant more importance to

what lies in the near future than to what lies in the distant future.

Psychoeducation: method used in clinical psychology which consists in explaining psychology and how the brain works in order to better understand how our behaviours, emotions and thoughts are formed.

Representativeness Bias: tendency to judge a person or a situation and to justify this judgement with a number of elements which are limited but which we consider typical of this person or situation.

Selection Bias: tendency, in the study of a subject, to select some pieces of information to the detriment of others that may be as relevant, which drives us to have a truncated vision of the subject matter in question.

Social Conformity: an attitude pushing us to adopt the same behaviour as a group of individuals (social class, political group, family sphere...).

Notes

1. S. L. Macknik, M. King, J. Randi, A. Robbins, Teller, J. Thompson and S. Martinez-Conde, 'Attention and awareness in stage magic: turning tricks into research', *Nature Reviews Neuroscience*, 9 (2008), p. 871–879.

2. J. Lehrer, 'Magic and the Brain: Teller Reveals the Neuroscience of Illusion', *Wired.com* (2009).

3. R. R. Trifiletti and E. H. Syed, 'Anton-Babinski Syndrome in a Child with Early-Stage Adrenoleukodystrophy', *European Journal of Neurology*, 14, n° 2 (2007).

4. E. F. Loftus and J. C. Palmer, 'Reconstruction of auto-mobile destruction: An example of the inter-action between language and memory', *Journal of Verbal Learning and Verbal Behavior*, 13, n° 5 (1974), p. 585-589.

5. https://www.innocenceproject.org.

6. E.F.Loftus, J.Coanet, J.E.Pickrell, 'Manufacturing false memories using bits of reality', in L. M. Reder, *Implicit Memory and Metacognition* (1996).

7. 'Grassement payée, la thérapeute faisait remonter de faux souvenirs', *Europe1.fr* (2017).

8. K. Abramson, 'Turning Up The Lights On Gaslighting', *Philosophical Perspectives*, 28, n° 1 (2014), p. 1-30.

9. P. Johansson, L. Hall, S. Sikström and A. Olsson, 'Failure to detect mismatches between intention and outcome in a simple decision task', *Science*, 310, n° 5745 (2005), p. 116-119.

10. A. Tversky and D. Kahneman, 'Judgment under Uncertainty: Heuristics and Biases', *Science*, 185, n° 4157 (1974), p. 1124-1131.

11. A. P. Gregg, N. Mahadevan, C. Sedikides, 'The SPOT effect: People spontaneously prefer their own theories', *The Quartely Journal of Experimental Psychology*, 70, n° 6 (2017).

12. *Ibid.*

13. J. Fox, 'Instinct Can Beat Analytical Thinking', *Harvard Business Review* (2014).

14. D. E. Melnikoff, J. A. Bargh, 'Trends in Cognitive Sciences', *Science Direct*, 22, n° 4 (2018).

15. D. E. Melnikoff and J. A. Bargh, 'The Mythical Number Two', Trends in Cognitive *Sciences*, 22, n° 4 (2018).

16. B.M. Galla, and A.L. Duckworth, 'More than resisting temptation: beneficial habits mediate the relationship between self-control and positive life outcomes', *Journal of Personality and Social Psychology*, 109 (2015), p. 508–525. W. Wood,

and D. Rünger, 'Psychology of habits', *Annual Review of Psychology*, 37 (2006), p. 289–314. A. Fishbach, L. Shen, 'The explicit and implicit ways of overcoming temptation' in J. W. Sherman, B. Gawronski and Y. Trope (Eds.), *Dual-process theories of the social mind*, Guilford Press, New York (2014) p. 454-467.

17. *Ibid.*

18. World Bank Group, 'World Development Report 2015: Mind, Society, and Behavior', *World Bank* (2015).

19. Agence européenne pour la sécurité and la santé au travail – EU-OSHA, 'Analyse documentaire : Calcul des coûts du stress et des risques psychosociaux liés au travail', *Publications Office of the European Union*, (2014). K. H. Pribram, 'A Review of Theory in Physiological Psychology', *Annual Review of Psychology*, 11 (1960), p. 1-40.

20. E.P. Balogh, B.T. Miller and R.B. Ball, 'Improving Diagnosis in Health Care', *National Academies Press*, Washington (2015). J. Hadwin, S. Frost, C. C. French, A. Richards, 'Cognitive processing and trait anxiety in typically developing children: Evidence for an interpretation bias', *Journal of Abnormal Psychology*, 106 n° 3 (1997), p. 486-490. M. R. Taghavi, A. R. Moradi, H. T. Neshat-Doost, W. Yule and Tim Dalgleish, 'Interpretation

of ambiguous emotional information in clinically anxious children and adolescents', *Cognition and Emotion*, 14, n° 6 (2010) p. 809-822 (2010). S. M. Bögels, D. Zigterman, 'Dysfunctional Cognitions in Children with Social Phobia, Separation Anxiety Disorder, and Generalized Anxiety Disorder', *Journal of Abnormal Child Psychology*, 28, n° 2 (2000), p. 205-211.

21. M. Spokas, R. G. Heimberg, T. Rodebaugh, 'Cognitive biases in social phobia', *Psychiatry*, 3, n° 5 (2004), p. 51-55.

22. W. S. Gilliam, A. N. Maupin, C. R. Reyes, M. Accavitti and F. Shic, 'Do Early Educators' Implicit Biases Regarding Sex and Race Relate to Behavior Expectations and Recommendations of Preschool Expulsions and Suspensions?', Yale University Child Study Center (2016).

23 A. Moukheiber, G. Rautureau, F. Perez-Diaz, R. Soussignan, S. Dubal, R. Jouvent and A. Pelissolo, 'Gaze avoidance in social phobia: objective measure and correlates', *Behaviour Research and Therapy*, 48, n° 2 (2010), p. 147-151.

24. B.K. Payne, 'Weapon Bias : Split-Second Decisions and Unintended Stereotyping', *Current directions in psychological science*, 15, n° 6, p. 291 (2006). J. Correll, B. Park, C.M. Judd, B. Wittenbrink, 'The police officer's dilemma: using ethnicity to

disambiguate potentially threatening individuals', *Journal of Personality and Social Psychology*, 83, n° 6 (2002), p. 1314-29.

25. D. Westen, P.S. Blagov, K. Harenski, C. Kilts, S. Hamann, 'Neural bases of motivated reasoning: an FMRI study of emotional constraints on partisan political judgment in the 2004 U.S. Presidential election', *Journal of Cognitive Neuroscience*, 18, n° 11 (2006), p. 1947-58.

26. J. Haidt, 'The Emotional Dog and Its Rational Tail: A Social Intuitionist Approach to Moral Judgment', *Psychological Review*, 108, n° 4 (2001), p. 814-834.

27. Y. Cahuzac and S. François, 'Les stratégies de communication de la mouvance identitaire. Le cas du Bloc identitaire', *Questions de communication*, 1, n° 23 (2013), p. 275-292.

28. C. Dovergne, 'Essena O'Neill, reine d'Instagram, raconte l'enfer derrière ses photos parfaites', *Vanity Fair* (2015).

29. L. Festinger, 'A Theory of cognitive dissonance', *Psychology coll.*, Stanford University Press, Stanford (1957).

30. B. Franklin, 'The Autobiography of Benjamin Franklin', *Americana coll.*, J.B. Lippincott & Co, Philadelphia (1868), p. 48.

31. J. Brehm, 'Postdecision Changes in the Desirability

of Alternatives', *Journal of Abnormal and Social Psychology*, 52, n° 3 (1956), p. 384-389. M. Chammar, I. E. Karoui, S. Allali, J. Hagège, K. Lehongre, D. Hasboun, M. Baulac, S. Epelbaum, A. Michon, B. Dubois, V. Navarro, M. Salti and L. Naccache, 'Cognitive Dissonance Resolution Depends on Episodic Memory', *Scientific Reports*, 7, n° 41320 (2017).

32. G. Russel, 'Le juteux business de l'indicateur de personnalité MBTI', *Le Figaro* (2004).

33. G. L. William and M. J. Martinko, 'Using the Myers-Briggs Type Indicator to Study Managers: A Literature Review and Research Agenda', *Journal of Management*, 22, n° 1, p. 45–83. D. J. Pittenger, 'Cautionary comments regarding the Myers-Briggs Type Indicator', *Consulting Psychology Journal: Practice and Research*, 57, n° 3 (2005), p. 210-221. R. Hogan, 'Personality and the fate of organizations', Lawrence Erlbaum Associates, New Jersey (2007), p. 28. W. L. Gardner and M. J. Martinko, 'Using the Myers-Briggs Type Indicator to Study Managers: A Literature Review and Research Agenda', *Journal of Management*, 22, n° 1 (2016), p. 45-83.

34. B. R. Forer, 'The Fallacy of personal validation: a classroom demonstration of gullibility', *The Journal of Abnormal and Social Psychology*, volume 44, n° 1, p. 118-123 (1949).

35. J.B. Rotter, 'Generalized expectancies for internal versus external control of reinforcement', *Psychological Monographs: General and Applied*, 80, n° 1 (1966), p. 1-28.

36. S. Jain and A. Pratap Singh, 'Locus of Control in Relation to Cognitive Complexity', *Journal of the Indian Academy of Applied Psychology*, 34, n° 1 (2008), p. 107-113.

37. E. J. Phares, 'Changes in expectancy in skill and chance situations', unpublished, doctoral dissertation, Ohio State University, Columbus (1955).

38. S. J. Spencer, C. M. Steele and D. M. Quinn, 'Stereotype Threat and Women's Math Performance', *Journal of Experimental Social Psychology*, 35, n° 1 (1999), p. 4-28.

39. N. Mamlin, K. R. Harris, L. P. Case, 'A Methodological Analysis of Research on Locus of Control and Learning Disabilities: Rethinking a Common Assumption', *Journal of Special Education*, 34, n° 4 (2001), p. 214-225.

40. *Ibid*.

41. J. M. Jacobs-Lawson, E. L. Waddell and A. K. Webb 'Predictors of Health Locus of Control in Older Adults', *Current Psychology*, 30, n° 2 (2011), p. 173-183.

42 M. E. P. Seligman, 'Learned Helplessness', *Annual Review of Medicine*, 23, n° 1 (1972), p. 407-412.

43. D. S. Hiroto, M. E. P. Seligman, 'Generality of Learned Helplessness in Man', *Journal of Personality and Social Psychology*, 31, n° 2 (1975), p. 311-327.

44. G. Ben-Shakhar, A. Y. Shalev and N. Bargai, 'Posttraumatic Stress Disorder and Depression in Battered Women: The Mediating Role of Learned Helplessness', *Journal of Family Violence*, 22, n° 5 (2007), p. 267-275.

45. E. Salomon, J. L. Preston and M. B. Tannenbaum, 'Climate Change Helplessness and the (De)moralization of Individual Energy Behavior', *Journal of Experimental Psychology Applied*, 23, n° 1 (2017), p. 1-13.

46 T. Griessinger, 'Apport des sciences comportementales aux politiques publiques pour la transition écologique', rapport d'étude, sous la direction de la DITP (2019).

47. S. Periasamy and J. S. Ashby, 'Multidimensional Perfectionism and Locus of Control, Adaptive vs. Maladaptive Perfectionism', *Journal of College Student Psychotherapy*, 17, n° 2 (2002), p. 75-86.

48. D. C. Watson, 'The Relationship of Self-Esteem, Locus of Control, and Dimensional Models to Personality Disorders', *Journal of Social Behavior and Personality*, 13, n° 3 (1998), p. 399.

49. M. A. Fuoco, 'Trial and error: They had larceny in their hearts, but little in their heads', *Pittsburgh Post-Gazette* (1996), p. D1.

50. J. Kruger and D. Dunning, 'Unskilled and Unaware of It: How Difficulties in Recognizing One's Own Incompetence Lead to Inflated Self-Assessments', *Journal of Personality and Social Psychology*, 77, n° 6 (1999), p. 1121-1134.

51. *Ibid.*

52. R. Lawson, 'The Science of Cycology: Failures to Understand How Everyday Objects Work', *Memory & Cognition*, 34, n° 8 (2006), p. 1667-1675.

53. B. Artz, A. H. Goodall and A. J. Oswald, 'Boss Competence and Worker Well-being', *Warwick Economics Research Paper Series*, n° 1072 (2015).

54. G. Wahl, *Les adultes surdoués,* Que sais-je, Presses Universitaires de France (2017), p. 65-74.

55. L. J. Peter and R. Hull, *The Peter Principle:Why Things Always Go Wrong*, HarperBusiness, (2011).

56. A. A. Durand, 'Les inégalités femmes-hommes en 12 chiffres and 6 graphiques', *LeMonde.fr* (2018).

57. P. Rose-Clance, *Le Complexe d'imposture*, Flammarion, (1986).

58. Source : www.insee.fr.

59. A. Boring, 'L'entrepreneuriat des femmes, objet de recherche à Science po', *EducPros.fr* (2017).

60. C. Gaubert, 'Vaccins : les Français reprennent confiance, d'après les industriels', *Sciencesetavenir. fr* (2018).

61. M. Motta, T. Callaghan and S. Sylvester, 'Knowing less but presuming more: Dunning-Kruger effects and the endorsement of anti-vaccine policy attitudes', *Social Science & Medicine*, 211 (2018), p. 274-281.

62. A. Hussain, S. Ali, M. Ahmed and S. Hussain, 'The Anti-vaccination Movement: A Regression in Modern Medicine', *Cureus*, 10, n° 7 (2018), p. e2919.

63. M. Fisher and M. Kranish, 'Trump Revealed: An American Journey of Ambition, Ego, Money and Power', Simon & Schuster (2016).

64. A. Selyukh, 'After Brexit Vote, Britain Asks Google: "What Is The EU?"', *npr.org* (2016).

65. '42% des Anglais pour un nouveau référendum', *Lematin. ch* (2019).

66. G. Pennycook, J. A. Cheyne, N. Barr, D. J. Koehler and J. A. Fugelsang, 'On the Reception and Detection of Pseudo-Profound Bullshit', *Judgment and Decision Making*, 10, n° 6 (2015), p. 549-563.

67. H. G. Frankfurt, *'On Bullshit'*, Princeton University Press (2005).

68. G. Pennycook, J. A. Cheyne, N. Barr, D. J. Koehler and J. A. Fugelsang, 'On the Reception

and Detection of Pseudo-Profound Bullshit', *Judgment and Decision Making*, 10, n° 6 (2015), p. 549-563.

69. S. R. Ketabi, 'Ayurvéda, le guide de référence', Guy Tredaniel (2018).

70. J. M. Darley and C. D Batson (1973). 'From Jerusalem to Jericho: A study of situational and dispositional variables in helping behavior', *Journal of Personality and Social Psychology*, 27, n° 1 (1973), p. 100-108.

71. T. Horanont, S. Phithakkitnukoon, T. W. Leong, Y. Sekimoto and R. Shibasaki, 'Weather Effects on the Patterns of People's Everyday Activities: A Study Using GPS Traces of Mobile Phone Users', *PLoS One*, 8, n° 12 (2013), p. e81153.

72. R. A. Baron, 'The Sweet Smell of... Helping: Effects of Pleasant Ambient Fragrance on Prosocial Behavior in Shopping Malls', *Personality and Social Psychology Bulletin*, 23, n° 5 (1997), p. 498-503.

73. L. Ross, 'The Intuitive Psychologist And His Shortcomings: Distortions in the Attribution Process', *Advances in Experimental Social Psychology*, 10 (1977) p. 173-220.

74. E. J. Johnson and D. Goldstein, 'Do Defaults Save Lives?', *Science Mag*, 302, n° 5649 (2003), p. 1338-1339.

75. Synthèse de presse bioéthique, 'Don d'organes : les allemands dépendants des autres pays européens', *genethique. org* (2018).

76. K. Moskvitch, 'The Road Design Tricks That Make Us Drive Safer', *BBC future* (2014).

77. P. Capelli, 'Les "nudges", force de persuasion', *Libération* (2014).

78 P. Gulborg Hansen and A. M. Jespersen, 'Nudge and the Manipulation of Choice, A Framework for the Responsible Use of the Nudge Approach to Behaviour Change in Public Policy', *European Journal of Risk Regulation*, 4, n° 1 (2013), p. 3-28.

79 S.E.Asch, 'Opinions and Social Pressure', *Scientific American*, 193, n° 5 (1955), p. 31-35.

80. M. M. Hollander and J. Turowetz, 'Normalizing trust: Participants' immediately post-hoc explanations of behaviour in Milgram's 'obedience' experiments', *The British Psychological Society*, 56, n° 4 (2017), p. 655-674. S. A. Haslam, S. D. Reicher, K. Millard and R. McDonald, '"Happy to have been of service":The Yale archive as a window into the engaged followership of participants in Milgram's 'obedience' experiments', *The British Psychological Society*, 54, n° 1 (2015), p. 55-83.

81. IIGEA, 'Sismo artificial por celebracion de gol en México.', *Sismologí* (2018).

82. J. Navajas, T. Niella, G. Garbulsky, B. Bahrami & M. Sigman, 'Aggregated knowledge from a small number of debates outperforms the wisdom of large crowds', *Nature Human Behaviour*, 2 (2018), p. 126-132.

83. J. M. Darley, B. Latané, 'Bystander intervention in emergencies: diffusion of responsibility', *Journal of Personality and Social Psychology*, 8, n° 4 (1968), p. 377–383.

84. D. M. Faris, 'La révolte en réseau: le "printemps arabe" and les médias sociaux', *Politique étrangère*, Printemps, n° 1 (2012), p. 99-109.

85. N. Normann, A. Emmerik, N. Morina, 'The Efficacy of Metacognitive Therapy for Anxiety and Depression: A Meta-Analytic Review', *Depression and Anxiety*, 31, n° 5 (2014), p. 402-411.

86. C. Sagan, 'The Demon-Haunted World: Science as a Candle in the Dark', Random House (1995).

87. S. Aral, S. Vosoughi and D. Roy, 'The Spread of True and False News Online', *Science*, 359, n° 6380 (2018), p. 1146-1151.

88. C. Levenson, 'En Inde, des rumeurs sur WhatsApp mènent au lynchage de sept hommes.', *Slate.fr* (2017).

89. G. Pennycook and D. G. Rand, 'Who falls for fake news? The roles of bullshit receptivity, over-claiming, familiarity, and analytic thinking', *Social Science Research Network* (2017).

90. L. Lamperouge, L. Mugiwara, I. Kurosaki, L. Cohen and D. Bowie, 'We do not tend to verify what we read' (2019). http://bit.ly/poneglyph.

91. S. Wineburg and S. McGrew 'Lateral Reading: Reading Less and Learning More When Evaluation Digital Information', *Stanford History Education Group Working*, n° 2017-A1 (2017).

92. AFP, 'Google, Facebook s'associent aux médias du "Trust Project"', *Le Point* (2017).

93 M. Farina, E. Pasquinelli, G. Zimmerman, *Esprit critique, esprit scientifique*, Éditions Le Pommier (2017).

94. E. R. Lai, *Critical Thinking: A Literature Review*, Parsons Publishing, (2011), p. 40-41.